ADVERTISING: MEDIA, MARKETING AND CONSUMER DEMANDS

COSMETICS AND FDA REGULATION

ADVERTISING: MEDIA, MARKETING AND CONSUMER DEMANDS

Additional books in this series can be found on Nova's website under the Series tab.

Additional E-books in this series can be found on Nova's website under the E-book tab.

GOVERNMENT PROCEDURES AND OPERATIONS

Additional books in this series can be found on Nova's website under the Series tab.

Additional E-books in this series can be found on Nova's website under the E-book tab.

ADVERTISING: MEDIA, MARKETING AND CONSUMER DEMANDS

COSMETICS AND FDA REGULATION

ADAM GARCIA
AND
ROSE DIBARTOLO
EDITORS

New York

For permission to use material from this book please contact us:
Telephone 631-231-7269; Fax 631-231-8175
Web Site: http://www.novapublishers.com

Additional color graphics may be available in the e-book version of this book.

Library of Congress Cataloging-in-Publication Data

ISBN: 978-1-62257-892-4

Published by Nova Science Publishers, Inc. † New York

CONTENTS

PREFACE

The 1938 Federal Food, Drug, and Cosmetic Act (FFDCA) granted the Food and Drug Administration (FDA) the authority to regulate cosmetic products and their ingredients. The statutory provisions of the FFDCA that address cosmetics include adulteration and misbranding provisions. Cosmetics are arguably more self-regulated than other FDA-regulated products. The manner in which a cosmetic product could or should be regulated, however, is not always clear. FDA's guidelines have provided the cosmetic industry with considerable flexibility for product development and claims. The question remains as to whether that flexibility and the extent of government oversight of cosmetic products are still appropriate. This book provides an overview of the cosmetics industry and FDA regulation.

Chapter 1 – The 1938 Federal Food, Drug, and Cosmetic Act (FFDCA) granted the Food and Drug Administration (FDA) the authority to regulate cosmetic products and their ingredients. The statutory provisions of the FFDCA that address cosmetics include adulteration and misbranding provisions. In addition to the FFDCA, cosmetics are regulated under the Fair Packaging and Labeling Act (FPLA) and related regulations. The cosmetics provisions were amended by the Color Additive Amendments Act of 1960 and the Poison Prevention Packaging Act, but remain basically the same as the provisions in the 1938 FFDCA.

FDA's authorities over cosmetic products include some of those applicable to other FDA-regulated products, such as food, drugs, medical devices, and tobacco. For example, FDA has the authority to take certain enforcement actions—such as seizures, injunctions, and criminal penalties—against adulterated or misbranded cosmetics. Additionally, as with drug and food companies, FDA may conduct inspections of cosmetic manufacturers and prohibit imports of cosmetics that violate the FFDCA. The agency also has issued rules restricting the use of ingredients that the agency has determined are poisonous or deleterious.

However, FDA's authority over cosmetics is less comprehensive than its authority over other FDA-regulated products with regard to registration; testing; premarket notification, clearance, or approval; good manufacturing practices; mandatory risk labeling; adverse event reports; and recalls. For example, FDA does not impose registration requirements on cosmetic manufacturers. Rather, cosmetic manufacturers may decide to comply with voluntary FDA regulations on registration. With the exception of color additives, FDA does not require premarket notification, safety testing, review, or approval of the chemicals used in cosmetic products. Cosmetic manufacturers also are not required to use good manufacturing practices (GMP)—although FDA has released GMP guidelines for cosmetic manufacturers—nor required to file ingredient information with, or report adverse reactions to, the agency. Instead, under a voluntary FDA program, cosmetic manufacturers and packagers may report the ingredients used in their product formulations. FDA does not have the authority to require a manufacturer to recall a cosmetic product from the marketplace, although the agency has issued general regulations on voluntary recalls. The agency's ability to issue regulations on cosmetic products is limited by the agency's statutory authorities or lack thereof.

As a result, cosmetics are arguably more self-regulated than other FDA-regulated products. The manner in which a cosmetic product could or should be regulated, however, is not always clear. FDA's guidelines have provided the cosmetic industry with considerable flexibility for product development and claims. The question remains as to whether that flexibility and the extent of government oversight of cosmetic products are still appropriate.

Chapter 2 – The cosmetics industry has been regulated by FDA since the enactment of the Federal Food, Drug and Cosmetic Act of 1938 (FFDCA). Currently, FDA's CFSAN is responsible for regulating cosmetics. Similar to drugs, devices and food, the FFDCA prohibits the introduction of adulterated or misbranded cosmetics into interstate commerce and provides for seizure, criminal penalties and other enforcement authorities for violations of the FFDCA. In addition, under the authority of the Fair Packaging and Labeling Act (FPLA), FDA requires an ingredient declaration for cosmetics to enable consumers to make informed purchasing decisions. Cosmetics that fail to comply with the FPLA are considered misbranded under the FFDCA.

Chapter 3 – This is Statement of Joseph R. Pitts.

Chapter 4 – This is Statement of Michael M. Landa.
Chapter 5 – This is Testimony of Halyna Breslawec.
Chapter 6 – This is Testimony of Peter Barton Hutt.
Chapter 7 – This is Testimony of Curran Dandurland.
Chapter 8 – This is Testimony of Deborah May.
Chapter 9 – This is Testimony of Dr. Michael DiBartolomeis.

In: Cosmetics and FDA Regulation
Editors: A. Garcia and R. DiBartolo
ISBN: 978-1-62257-892-4

Chapter 1

FDA REGULATION OF COSMETICS AND PERSONAL CARE PRODUCTS*

Amalia K. Corby-Edwards

SUMMARY

The 1938 Federal Food, Drug, and Cosmetic Act (FFDCA) granted the Food and Drug Administration (FDA) the authority to regulate cosmetic products and their ingredients. The statutory provisions of the FFDCA that address cosmetics include adulteration and misbranding provisions. In addition to the FFDCA, cosmetics are regulated under the Fair Packaging and Labeling Act (FPLA) and related regulations. The cosmetics provisions were amended by the Color Additive Amendments Act of 1960 and the Poison Prevention Packaging Act, but remain basically the same as the provisions in the 1938 FFDCA.

FDA's authorities over cosmetic products include some of those applicable to other FDA-regulated products, such as food, drugs, medical devices, and tobacco. For example, FDA has the authority to take certain enforcement actions—such as seizures, injunctions, and criminal penalties—against adulterated or misbranded cosmetics. Additionally, as with drug and food companies, FDA may conduct inspections of cosmetic manufacturers and prohibit imports of cosmetics that violate the FFDCA. The agency also has issued rules restricting the use of ingredients that the agency has determined are poisonous or deleterious.

* This is an edited, reformatted and augmented version of Congressional Research Service, Publication No. R42594, dated July 9, 2012.

However, FDA's authority over cosmetics is less comprehensive than its authority over other FDA-regulated products with regard to registration; testing; premarket notification, clearance, or approval; good manufacturing practices; mandatory risk labeling; adverse event reports; and recalls. For example, FDA does not impose registration requirements on cosmetic manufacturers. Rather, cosmetic manufacturers may decide to comply with voluntary FDA regulations on registration. With the exception of color additives, FDA does not require premarket notification, safety testing, review, or approval of the chemicals used in cosmetic products. Cosmetic manufacturers also are not required to use good manufacturing practices (GMP)—although FDA has released GMP guidelines for cosmetic manufacturers—nor required to file ingredient information with, or report adverse reactions to, the agency. Instead, under a voluntary FDA program, cosmetic manufacturers and packagers may report the ingredients used in their product formulations. FDA does not have the authority to require a manufacturer to recall a cosmetic product from the marketplace, although the agency has issued general regulations on voluntary recalls. The agency's ability to issue regulations on cosmetic products is limited by the agency's statutory authorities or lack thereof.

As a result, cosmetics are arguably more self-regulated than other FDA-regulated products. The manner in which a cosmetic product could or should be regulated, however, is not always clear. FDA's guidelines have provided the cosmetic industry with considerable flexibility for product development and claims. The question remains as to whether that flexibility and the extent of government oversight of cosmetic products are still appropriate.

INTRODUCTION

The U.S. cosmetic, beauty supply, and perfume retail industry consists of approximately 13,000 establishments, with annual revenue of about $10 billion.[1] Worldwide, the cosmetics and personal care products industry has more than $250 billion in annual retail sales.[2] According to economic census data released in 2009, the U.S. cosmetic industry employs over 86,000 people.[3]

The cosmetic market includes numerous personal care products that have many uses beyond the facial makeup that one typically thinks of when the term "cosmetics" is used. Industry sales are concentrated in the following areas (percentage of sales by product category): (1) cosmetics, face cream, and perfume—75%; (2) hygienic products including deodorant, shampoo, conditioner, hair color, and shaving products—20%; and (3) small appliances—4%.[4] The typical industry consumer is a woman between the ages of 25 to 55, although there appears to be increasing growth in marketing to men and tweens (9- to 12-year-olds).[5] Sales of cosmetic and personal care products may be affected by a consumer's personal income, although the "sales of basic personal items such as soap, shampoo, and shaving products are likely to be less impacted by a soft economy than other product areas viewed by consumers as more discretionary."[6] Prices for cosmetics vary widely, and depend on whether the product is a "prestige," mass market, or a professional or salon use brand.[7]

The Food and Drug Administration (FDA) reportedly regulates $62 billion worth of cosmetics.[8] FDA's primary responsibilities for regulating cosmetics include ensuring that cosmetics are not adulterated or misbranded.[9] This chapter describes the differences between cosmetics, drugs, and combination products; provides an overview of the statutory provisions and rules under which FDA regulates cosmetics; and provides an overview of industry self-regulation programs. The chapter also includes an appendix on keratin hair treatment products, also known as "Brazilian Blowouts." This chapter focuses on FDA regulation of cosmetics and does not discuss Federal Trade Commission regulation of advertising of cosmetics nor the regulation of potentially dangerous chemicals or pesticides by other agencies, with the exception of formaldehyde and other agents that may produce or lead to the production of formaldehyde.[10]

COSMETICS, DRUGS, AND COMBINATION PRODUCTS

This section discusses the Federal Food, Drug, and Cosmetic Act (FFDCA) definitions of cosmetics and drugs, and how the FFDCA differentiates between cosmetics and a cosmetic that also meets the statutory definition of a drug. Classification of products is a concern for manufacturers, as cosmetics are not subject to the same approval, regulatory, or registration requirements as drugs.[11]

In addition to saving considerable time and expense, this distinction allows manufacturers of products that are only cosmetics and not drugs or combination products, discussed later, to market their products with less regulatory oversight.

Cosmetics

The term "cosmetics" covers a broad range of FDA-regulated products that may be used externally, orificially, and internally.[12] For regulatory purposes, the term "cosmetics" includes products for the eyes, face, nails, hair, skin, and mouth, which may be in the form of products such as makeup, polish, hair dyes and coloring, sunscreens, fragrances, shave gel, oral care and bath products, and products for infants and children.[13] In some settings, cosmetics are known as "personal care products" because of the wide range of products now regulated as cosmetics that are not strictly facial cosmetics. For purposes of this chapter, "cosmetics" will be used to refer to the entire category of products being discussed.

The FFDCA defines "cosmetics" as "(1) articles intended to be rubbed, poured, sprinkled or sprayed on, introduced into, or otherwise applied to the human body or any part thereof for cleansing, beautifying, promoting attractiveness, or altering the appearance, and (2) articles intended for use as a component of any such articles; except that the term shall not include soap."[14] While soap was explicitly exempted from the definition of a cosmetic, and is not defined in the FFDCA, it is defined in FDA regulations.[15] Additionally, coal tar hair dye was provided a limited exemption from the FFDCA's adulteration provisions.[16]

Drugs

The FFDCA defines a "drug" as including articles "intended for use in the diagnosis, cure, mitigation, treatment, or prevention of disease," articles that are "intended to affect the structure or any function of the body," and "articles intended for use as a component" of such drugs.[17]

Unlike cosmetics and their ingredients (with the exception of color additives), drugs are subject to FDA approval before they can enter interstate commerce. Drugs must either receive the agency's premarket approval of a new drug application[18] or conform to a set of FDA regulations known as a monograph. Monographs govern the manufacture and marketing of over-the-counter (OTC) drugs and specify the conditions under which OTC drugs in a particular category (such as antidandruff shampoos or antiperspirants) will be considered to be generally recognized as safe and effective.[19] Monographs also indicate how OTC drugs must be labeled so they are not deemed to be misbranded.[20] Such labeling includes a Drug Facts panel, which provides a listing of the active ingredients in the product as well as the drug's purposes, uses, and applicable warnings, directions, inactive ingredients, other information, and a telephone number for questions about the product.[21]

Drug manufacturers must comply with good manufacturing practices (GMP) rules for drugs; failure to follow GMP may cause a drug to be considered adulterated.[22] Drug manufacturers also are required to register their facilities, list their drug products with the agency, and report adverse events to FDA.[23]

Cosmetics Containing Drug Ingredients

While reference to "cosmetic drugs" or "cosmeceuticals" has been used by some proponents in referring to combination cosmetic-drug products, there is not an FDA statutory or regulatory definition for this terminology.[24] Cosmetic-drug combination products are subject to FDA's regulations for both cosmetics and drugs.

Comparison of Cosmetic and Drug Product Classifications
A suntan product is a cosmetic, but a sunscreen product is a drug. A deodorant is a cosmetic, but an antiperspirant is a drug. A shampoo is a cosmetic, but an antidandruff shampoo is a drug. A toothpaste is a cosmetic, but an anticavity toothpaste is a drug. A skin exfoliant is a cosmetic, but a skin peel is a drug. A mouthwash is a cosmetic, but an antigingivitis mouthwash is a drug. A hair bulking product is a cosmetic, but a hair growth product is a drug. A skin product to hide acne is a cosmetic, but an antiacne product is a drug. An antibacterial deodorant soap is a cosmetic, but an antibacterial anti-infective soap is a drug. A skin moisturizer is a cosmetic, but a wrinkle remover is a drug. A lip softener is a cosmetic, but a product for chapped lips is a drug.

Source: Peter Barton Hutt, "Legal Distinction in USA between Cosmetic and Drug," in *Cosmeceuticals and Active Cosmetics: Drugs versus Cosmetics,* p. 630 (Peter Elsner & Howard Maibach, eds., 2nd ed. 2005).

Combination drug and cosmetic products must meet both OTC drug and cosmetic labeling requirements, that is, the drug ingredients must be listed alphabetically as "Active Ingredients," followed by cosmetic ingredients either listed in a descending order of predominance as "Inactive Ingredients" or listed as "Inactive Ingredients" in particular groups, such as concentrations of greater than one percent of color additives.[25]

The determination of whether a cosmetic is also a drug, and therefore subject to the additional statutory requirements that apply to drugs, is based on the distributor's intent or the intended use.[26] The intended use of a product may be established in several ways, such as claims on the labeling or in advertising or promotional materials, or through the inclusion of ingredients that will cause the product to be considered a drug because of a known therapeutic use. For example, if a lipstick (a cosmetic) contains sunscreen (a drug), the mere inclusion of the term "sunscreen" in the product's labeling will cause the product to also be regulated as a drug.[27] The text box below provides examples of other combination products and compares cosmetic versus drug classifications.

OVERVIEW OF FDA'S AUTHORITY TO REGULATE COSMETICS

The FFDCA prohibits the adulteration and misbranding of cosmetics and the introduction, receipt, and delivery of adulterated or misbranded cosmetics into interstate commerce.[28] A cosmetic is considered to be adulterated if, among other reasons, it contains a substance which may cause injury to users under the conditions of use prescribed on the product's labeling or if it contains a filthy, putrid, or decomposed substance.[29] A cosmetic is considered to be misbranded if its labeling is false or misleading, if it does not bear the required labeling information, or if the container is made or filled in a deceptive manner.[30]

Prior to the enactment of the FFDCA in 1938, cosmetics were not regulated by the federal government,[31] but were regulated under a collection of state laws that had been enacted to regulate food and drugs.[32] At that time, several "cosmetics and drugs were made from the same natural materials" and the "laws did not include explicit definitions of the products regulated."[33] Following several incidents in which cosmetics were allegedly the cause of serious health problems, as well as industry concerns about states enacting their own laws, provisions were included in FFDCA that prohibited the sale of adulterated or misbranded cosmetics in interstate commerce.[34] The FFDCA also established uniform regulation of FDA-regulated cosmetic products throughout the country.[35]

In addition to the FFDCA, cosmetics are regulated under the Fair Packaging and Labeling Act (FPLA) and related regulations.[36] The FPLA applies to the packaging and labeling of "consumer commodities," which include cosmetics "customarily produced or distributed for sale through retail sales agencies or instrumentalities for consumption by individuals, or use by individuals for purposes of personal care ... and which [are] usually consumed or expended in the course of such consumption or use."[37] For the purposes of "for professional use only" labeling, discussed later, the FPLA does not apply to "wholesale or retail distributors of consumer commodities, except to the extent that such persons (1) are engaged in the packaging or labeling of such commodities, or (2) prescribe or specify ... the manner in which such commodities are packaged or labeled."[38]

The FFDCA statutory provisions that address cosmetics, with the exception of those regarding color additives, have remained basically unchanged since 1938, although the cosmetic industry today encompasses a greater number of products with different uses than those on the market more than seventy years ago. However, concerns of consumer and industry groups today are similar to those expressed prior to the enactment of the FFDCA. Consumer groups have raised concerns about particular ingredients, and states have considered legislating in areas not covered by the FFDCA or federal regulations.[39]

If a cosmetic that is introduced into, in, or held for sale after shipment in interstate commerce is found to be adulterated or misbranded, FDA may take enforcement actions, such as seeking an injunction (which could prevent a company from making or distributing the violative product), seizing the violative product, or seeking criminal penalties.[40] Additionally, FDA has authority to prevent imports of violative cosmetic products from entering the United States.[41]

FDA's authority to regulate cosmetics also includes the authority to conduct inspections of cosmetic establishments, without notifying the establishments in advance, as long as the inspections occur "at reasonable times and within reasonable limits and in a reasonable manner."[42] FDA conducts inspections to assure product safety and to evaluate cosmetic products for potential adulteration or misbranding violations.[43] The agency may decide to inspect a facility based on consumer or industry complaints, the establishment's compliance history, or FDA surveillance initiatives.[44] The agency may collect samples for examination and analysis during plant and import inspections, and follow up on complaints of adverse events alleged to be caused by a given cosmetic product.[45] The agency does not have a required schedule for inspecting cosmetic facilities.

FDA has certain regulations and procedures for cosmetics with which manufacturers voluntarily may choose to comply, even though similar regulations and procedures are mandatory for other FDA-regulated products. For example, FDA has regulations on voluntary facility registration and voluntarily reporting for ingredients used in cosmetic products and adverse reactions to cosmetics.[46] In contrast, registration requirements exist for other FDA product manufacturers.[47] Additionally, cosmetic manufacturers are not required, as drug manufacturers are, to "file data on ingredients, or report cosmetic-related injuries to FDA."[48] Instead, under a voluntary FDA program, cosmetic manufacturers and packagers may report the ingredients used in their product formulations.[49] Furthermore, consumers and cosmetic manufacturers may voluntarily report adverse reactions to cosmetics to FDA. Finally, FDA does not have mandatory recall authority to require a cosmetic manufacturer to recall a product from the marketplace. However, the agency may request a voluntary recall, and FDA has issued general regulations on the conduct of voluntary recalls that outline the agency's expectations of manufacturers during a recall.[50] While FDA does not have the authority to require compliance with these regulations, FDA may take action against adulterated or misbranded cosmetics.[51]

FDA's authority over cosmetics is less comprehensive than its authority over other FDA-regulated products with regard to GMP; premarket notification, clearance, or approval; testing; and mandatory risk labeling.[52] As an example, cosmetic producers are not required to use GMP unless their cosmetics are also drugs. FDA has released GMP guidelines for cosmetic manufacturers,[53] and has stated that "[f]ailure to adhere to GMP may result in an adulterated or misbranded product."[54] With the exception of color additives, FDA does not require premarket notification, safety testing, or premarket review or approval of the chemicals used in cosmetic products.[55] Also, unlike drugs, cosmetic products are not required to meet FDA requirements for safety and effectiveness.[56]

ADULTERATED AND MISBRANDED COSMETICS

As previously noted, the FFDCA prohibits the adulteration or misbranding of cosmetics, and the introduction, receipt, and delivery of adulterated or misbranded cosmetics into interstate commerce.[57] If a cosmetic that is introduced into, in, or held for sale after shipment in interstate commerce is found to be adulterated or misbranded, FDA may take enforcement actions. The following sections describe the parameters of the adulteration and misbranding of cosmetics.

Adulteration

A cosmetic is deemed adulterated—and potentially may be subject to FDA enforcement actions— if it

- "bears or contains any poisonous or deleterious substance which may render it injurious to users under the conditions of use prescribed in the labeling";
- consists of "any filthy, putrid, or decomposed substance";
- was "prepared, packed, or held under insanitary conditions whereby it may have become contaminated" or "rendered injurious to health";
- is in a container composed of "any poisonous or deleterious substance which may render the contents injurious to health"; or
- contains an unsafe color additive, except for hair dyes.[58]

FDA has issued rules restricting the use of some ingredients in cosmetic products, such as those that it has determined are poisonous or deleterious, which would cause the cosmetic to be adulterated.[59] One example of an adulterated cosmetic is the use of henna for a temporary skin decoration known as mehndi.[60] While the color additive used in these products is approved for hair dye, it is not permitted for skin contact.[61] Therefore, under FDA regulations, the use of the dye product in mehndi makes the product "adulterated."[62]

Misbranding and Mislabeling Claims

Cosmetic products that do not comply with FPLA requirements are considered misbranded under FFDCA, if they meet the FPLA's definition of "consumer commodities," discussed below. 63 64 Additionally, under FFDCA, cosmetics will be deemed to be misbranded, if

- the "labeling is false or misleading in any particular";
- the label lacks required information;[65]
- required labeling information is not prominently placed with conspicuousness and "in such terms as to render it likely to be read and understood by the ordinary individual under customary conditions of purchase and use";
- the "container is so made, formed, or filled as to be misleading";
- use of a color additive does not conform to packaging and labeling requirements; or
- the packaging or labeling violates the regulations issued under the Poison Prevention Packaging Act of 1970.[66]

Consumer commodity (retail) cosmetic products subject to the FPLA are required to bear a label with the identity of the product and the name and place of business of the manufacturer, packer, or distributor, as well as the net quantity of contents on the label's principal display panel.[67] The net quantity of contents information on a package's label must be declared in a legible type size that is uniform for packages of about the same size.[68] FDA's ingredient labeling rules, issued under the authority of the FPLA, require ingredients to be listed on cosmetic products in descending order of predominance.[69]

Enforcement

Consumer organizations and interested persons may submit citizen petitions to FDA asking the agency to determine that a cosmetic is adulterated if it contains a particular deleterious substance.[70] For example, in 1996, FDA denied such a petition after conducting a review of the cosmetic ingredient urocanic acid and "conclud[ing] that the scientific evidence did not establish urocanic acid to be a deleterious substance."[71]

If a cosmetic is deemed adulterated or misbranded, FDA may take enforcement actions. Enforcement actions may include seeking an injunction (which could prevent a company from making or distributing the violative product), seizing the violative product, or seeking criminal penalties.[72] Additionally, a cosmetic company may be subject to a product liability lawsuit for a product that could be deemed to be adulterated, misbranded, or that lacks adequate warning statements.[73]

VOLUNTARY RECALLS

FDA does not have authority to order a mandatory recall of a cosmetic product. In contrast, the agency has the authority to order recalls of food,[74] infant formula,[75] medical devices,[76] human tissue products,[77] and tobacco products.[78] Even though FDA may not order a mandatory recall, FDA may request that a company voluntarily recall cosmetic products.[79] Manufacturers or distributors may undertake voluntary recalls to remove violative products from the market that are hazardous to health, defective, or grossly deceptive, and "against which the agency would initiate legal action."[80] If a manufacturer or distributor is unwilling to remove dangerous products from the market without FDA's written request to do so, the agency may issue a request for a product recall.[81]

The agency monitors a firm that conducts a product recall, and the agency may take an active role in monitoring a recall by reviewing the firm's status reports and conducting its own audit checks to verify the recall's effectiveness.[82] FDA evaluates the health hazard presented by the product and assigns a classification to indicate the degree of hazard posed by the product under recall, whether it is a cosmetic or another FDA-regulated product (see text box).[83] Either FDA or the cosmetic company will issue public notification of the recall.[84] The firm is responsible for the disposition of the recalled product, whether it is destroyed or brought into compliance.[85]

Classification of Recall by Degree of Health Hazard

FDA evaluates the health hazard presented by the product and assigns a classification to indicate the degree of hazard posed by the product under recall.

Class I is a situation in which there is a reasonable probability that the use of, or exposure to, a violative product will cause adverse health consequences or death.

Class II is a situation in which use of, or exposure to, a violative product may cause temporary or medically reversible adverse health consequences or where the probability of serious adverse health consequences is remote.

Class III is a situation in which use of, or exposure to, a violative product is not likely to cause adverse health consequences.

Source: 21 C.F.R. § 7.3(m).

PREMARKET APPROVAL

In contrast to FDA's authority over drugs and some devices, FDA does not have the authority to require premarket approval of cosmetics or their ingredients, except for color additives.[86] Because there are no statutory requirements for premarket approval of cosmetic ingredients, manufacturers arc rcsponsible for substantiating the safety of their products and ingredients before the products are marketed.[87] Failure to adequately substantiate the safety prior to marketing causes the product to be considered misbranded, unless it bears a warning label that states: "The safety of this product has not been determined."[88] However, because that warning label seems to be rarely used, consumers may be under the impression that cosmetics have been demonstrated to be safe.[89] The Government Accountability Office (GAO) has noted that FDA's regulation requiring warning labels "cannot be effectively enforced because FDA does not have the authority to require cosmetic manufacturers to test their products for safety or make their test results available to FDA."[90]

FDA, however, has restricted the use of certain ingredients in cosmetics or required warning statements on the labels of certain types of cosmetics (see textbox below). For example, FDA issued a rule banning the use of methylene chloride in cosmetics after concluding "that methylene chloride is a poisonous or deleterious substance that may render cosmetic products injurious to users," due to the potential cancer risks of exposure to the substance.[91] If a cosmetic were to contain methylene chloride, it would be considered adulterated, and FDA could take an enforcement action.

Except for color additives and those cosmetic ingredients that are prohibited or restricted for use by a specific regulation, any ingredient used in the formulations of cosmetics is allowed, provided that the safety of the ingredient has been adequately substantiated, it is properly labeled, and its use does not cause the product to be adulterated or misbranded under the law. FDA's guidance document on inspections of cosmetic product manufacturers discusses several other ingredients that investigators should document if they are used in cosmetic products.[92]

Testing and Safety of Cosmetic Ingredients

FDA has advised cosmetic firms to employ appropriate and effective testing to substantiate the safety of their products.[93] However, the FFDCA does not specify how cosmetic products and their ingredients are to be tested.[94] As mentioned earlier, manufacturers are responsible for substantiating the safety of both the ingredients and finished cosmetic products prior to marketing.[95]

FDA Regulations for Certain Cosmetic Ingredients or Products

FDA has either restricted the use of the following ingredients in cosmetics or required warning statements on the labels of certain types of cosmetics.

21 C.F.R. § 250.250 Hexachlorophene

21 C.F.R. § 700.11 Cosmetics containing bithionol

21 C.F.R. § 700.13 Use of mercury compounds in cosmetics including use as skinbleaching agents in cosmetic reparations also regarded as drugs

21 C.F.R. § 700.14 Use of vinyl chloride as an ingredient including propellant of cosmetic aerosol products

21 C.F.R. § 700.15 Use of certain halogenated salicylanilides as ingredients in cosmetic products

21 C.F.R. § 700.16 Use of aerosol cosmetic products containing zirconium

21 C.F.R. § 700.18 Use of chloroform as an ingredient in cosmetic products

21 C.F.R. § 700.19 Use of methylene chloride as an ingredient of cosmetic products

21 C.F.R. § 700.23 Chlorofluorocarbon propellants

21 C.F.R. § 700.27 Use of prohibited cattle materials in cosmetic products

21 C.F.R. § 700.35 Cosmetics containing sunscreen ingredients

21 C.F.R. § 740.10 Labeling of cosmetic products for which adequate substantiation of safety has not been obtained

21 C.F.R. § 740.11 Cosmetics in self-pressurized containers

21 C.F.R. § 740.12 Feminine deodorant sprays

21 C.F.R. § 740.17 Foaming detergent bath products

21 C.F.R. § 740.19 Suntanning preparations

Source: Title 21, Code of Federal Regulations.

Traditional testing of cosmetic ingredients has used animal models to evaluate the safety of the ingredients on the human body. The tests used historically include measures of skin irritancy, eye irritation, allergic reactions, and toxicity caused by various ingredients used in the manufacture of cosmetics on several different animals, including rabbits, mice, rats, and guinea pigs.[96] Animal testing is allowed to be used to establish product safety.[97]

While concerns about the safety of cosmetics have been raised over the years, animal rights advocates have sought an end to animal testing.[98] FDA has said that it follows applicable laws on animal testing, such as the Animal Welfare Act.[99] Additionally, the agency has outlined its support for alternatives to whole-animal testing:

> FDA supports and adheres to the provisions of applicable laws, regulations, and policies governing animal testing, including the Animal Welfare Act and the Public Health Service Policy of Humane Care and Use of Laboratory Animals. Moreover, in all cases where animal testing is used, FDA advocates that research and testing derive the maximum amount of useful scientific information from the minimum number of animals and employ the most humane methods available within the limits of scientific capability ... We also believe that prior to use of animals, consideration should be given to the use of scientifically valid alternative methods to whole-animal testing. ...
>
> FDA supports the development and use of alternatives to whole-animal testing as well as adherence to the most humane methods available within the limits of scientific capability when animals are used for testing the safety of cosmetic products. We will continue to be a strong advocate of methodologies for the refinement, reduction, and replacement of animal tests with alternative methodologies that do not employ the use of animals.[100]

Cosmetic Ingredient Review Program

Although the FFDCA does not specify how ingredients in cosmetic products are to be tested, the cosmetic industry's trade association—the Personal Care Products Council (PCPC)—has established a Cosmetic Ingredient Review (CIR) program to review the safety of cosmetic product ingredients, based on published and unpublished data on individual ingredients. The purpose of the CIR program "is to determine those cosmetic ingredients for which there is a reasonable certainty in the judgment of competent scientists that the ingredient is safe under its conditions of use."[101]

Under the CIR program, an expert panel reviews cosmetic ingredients based on an annual priority list of ingredients currently used in commercially available cosmetics, which is based upon "the number of different products in which an ingredient is used" as obtained from the Voluntary Cosmetic Registration Program[102] as well as "toxicological considerations."[103] Panelists analyze data and determine whether an ingredient is (1) safe for the uses and concentrations in the safety assessment; (2) unsafe and therefore unsuitable for use in cosmetics; (3) safe, with qualifications, as in it can be used under certain conditions; or (4) an ingredient for which data are insufficient.[104] Although CIR's ingredient findings are published, the cosmetic industry is not required to follow CIR findings.[105]

As of February 2012, CIR has determined 1,398 ingredients "safe as used";[106] 987 ingredients safe with qualifications;[107] 43 ingredients with insufficient data to support safety;[108] and 11 ingredients "unsafe for use in cosmetic products."[109]

In addition, the Research Institute for Fragrance Materials (RIFM) "conducts a companion program to review the safety of fragrance ingredients" that includes a "systematic evaluation of fragrance ingredients used in cosmetic products."[110]

Consumer Concerns About the Safety of Ingredients

In 2004, concerns raised about the safety of some cosmetics led to the creation of a national coalition of environmental, health, labor, consumer, and women's groups called the Campaign for Safe Cosmetics.[111] The Campaign is concerned about what it believes to be a growing body of evidence that suggests a connection between certain chemicals and long-term health effects such as cancer and reproductive problems. Of particular concern are the health effects of nitrosamines,[112] lead and other heavy metals,[113] parabens,[114] phthalates,[115] hydroquinone,[116] and 1,4-dioxane.[117]

Beginning in 2004, the Campaign asked cosmetic companies to sign the Compact for Safe Cosmetics, which was a voluntary pledge by companies to take steps including disclosure of all ingredients, publication of product information in an ingredient database, and substantiation of "the safety of all products and ingredients with publicly available data."[118] More than 1,500 companies have signed the pledge to remove hazardous chemicals and replace them with safe alternatives within three years.[119]

The Campaign for Safe Cosmetics has also issued reports, which cover subjects such as contaminants in children's bath and personal care products.[120] The Environmental Working Group—a member of the Campaign for Safe Cosmetics—maintains a database of cosmetic product ingredients and related safety information.[121]

CONCERNS ABOUT SPECIFIC INGREDIENTS

Some ingredients used in cosmetic products have received particular attention as a result of concerns about their potential health risk. For example, questions have been raised about the accuracy of ingredient statements and the adequacy of safety warnings on product labels for keratin hair treatment products[122] containing formaldehyde.[123] Concerns have also arisen regarding the use of coal tar hair dyes as color additives [124] and nanomaterial ingredients, which are discussed more below.

Color Additives

As previously discussed, FDA does not require premarket approval of cosmetic ingredients, except for color additives. FDA regulates color additives—such as FD&C Blue No. 1— differently than other cosmetic ingredients and differently for use in cosmetics than for use in food, drugs, or medical devices. Color additives include any dye, pigment, or substance that may impart a color when added to a food, drug, cosmetic, or the human body,[125] and must be listed in a regulation before they are allowed to be used.[126] A cosmetic that contains a color additive that does not comply with the applicable FDA regulation will cause the cosmetic product to be considered to be adulterated.[127]

Additionally, some color additives must be certified by FDA before they may be used.[128] Failure to certify a color additive may cause the entire cosmetic product in which it is used to be deemed to be adulterated.[129] Batches of color additives are either subject to, or exempt from, certification by FDA.[130] The color additives that are subject to certification "are derived primarily from petroleum," while color additives exempt from certification "are obtained primarily from mineral, plant, or animal sources."[131] Regardless of whether a color additive is subject to certification, all color additives must be approved as "safe-for-use" prior to being listed and therefore able to be used in cosmetics.[132]

In addition to being subject to certification by FDA, color additives must be used according to FDA regulations that prescribe "the conditions under which such additive may be safely used."[133] For example, the color additive FD&C Red No. 4 must meet the requirements of 21 C.F.R. § 74.1304(a)(1) and (b), which discuss identity (the composition and specifications the color additive must meet, such as the maximum amounts of particular impurities that the color additive can contain) and restrict its use to "externally applied drugs and cosmetics."[134] Under FDA regulations, the external application of cosmetics does not include "the lips or any body surface covered by mucous membrane," and therefore FDA regulations prohibit the use of certain colors in cosmetics such as lipsticks.[135] As additional examples, FDA has specific regulations for an approved glow-in-the-dark color additive and for fluorescent color additives (some of which are approved for use in cosmetics) and for liquid crystal color additives (which are unapproved color additives and, therefore, are not approved for use in cosmetics).[136] FDA regulations also contain restrictions on color additives for use in the eye area, in injections (such as for tattoos or permanent makeup), and in surgical sutures, including that the listing or certification of the color additive must allow that specific use.[137]

Coal Tar Hair Dyes

Coal tar dyes have been a particularly controversial group of color additives, due to their potential health risk. Coal tar dyes are "synthetic-organic" colors, most of which are "made from petroleum."[138] These dyes, "which deposit and adhere to the hair shaft," "are either listed and certified color additives or dyes for which approval has not been sought."[139] They were specifically exempted from the FFDCA adulteration and other color additive provisions for products that are intended to dye hair.[140]

FDA, GAO, policymakers, and consumer groups have questioned whether the FFDCA exemption for coal tar hair dyes should be repealed because of potential health hazards.[141] On several occasions, the FDA unsuccessfully has argued for the repeal of the coal tar hair dye exemption.[142] The GAO also "recommended that FDA evaluate safety data on coal tar hair dye ingredients and require, where applicable, a cancer or other appropriate warning statement on product labels."[143] FDA has stated that "several coal-tar hair dye ingredients have been found to cause cancer in laboratory animals."[144] FDA unsuccessfully attempted to require the following warning on hair dyes that contained the coal tar ingredient 4-methoxy-m-phenylenediamine (4-MMPD, 2, 4- diaminoanisole): "Warning—Contains an ingredient that can penetrate your skin and has been determined to cause cancer in laboratory animals."[145]

Coal tar dyes are explicitly excluded from use in products intended to be dyes for eyelashes or eyebrows.[146] To avoid an adulteration determination, coal tar hair dyes must contain the FFDCAmandated warning statement that informs consumers of the potential risks associated with their use: "Caution – This product contains ingredients which may cause skin irritation on certain individuals and a preliminary test according to accompanying directions should first be made. This product must not be used for dyeing the eyelashes or the eyebrows; to do so may cause blindness."[147]

Nanomaterial Ingredients

The inclusion of nanomaterial ingredients in cosmetics has generated debate over the safety of nanomaterials and how and whether FDA should regulate such ingredients. Nanotechnology involves the application and manipulation of small matter "at the nanoscale, which is about 1 to 100 nanometers."[148] The cosmetic industry has used nanotechnology in cosmetic products for more than two decades.[149] Cosmetics are reportedly "the most prominent nanotechnology products on the U.S. market,"[150] and the "global market for cosmetics using nanotechnology [was] projected to reach an estimated $155.8 [million] in 2010."[151] Nanomaterials are reportedly used in two main ways in cosmetic products—as UV filters and as delivery systems.[152] The Project on Emerging Nanotechnologies—created in 2005 as a partnership between the Pew Charitable Trusts and the Woodrow Wilson International Center for Scholars—maintains a searchable database of consumer products, including cosmetics, that reportedly contain nanomaterials.[153] Cosmetic products with nanomaterials include facial cosmetic products, from creams and moisturizers to bronzers and blushers to mascara.[154]

There is debate among the scientific community as to the potential health effects of these particles. In general, the concerns about the use of nanomaterials in FDA-regulated products surround whether the small size of these particles leads to any new toxicological properties or harmful health effects, such as potentially damaging the skin or "crossing into the bloodstream, cells, and organs."[155] The unique size and chemical properties of these materials has led to concerns that they may have an increased ability to permeate the human skin and may release toxins into the bloodstream.[156] Damaged skin may be "especially at risk for nanoparticle penetration."[157] Other issues may include access to the body by inhalation, ingestion, or skin penetration; the length of time that they remain in the body; the dose likely to cause harm; the effects of long-term exposure; and the impact on the environment.[158] Consumer groups such as Friends of the Earth, the International Center for Technology Assessment, and Consumers Union have raised concerns about nanomaterials in cosmetic products and have petitioned FDA regarding the regulation of products containing nanomaterials.[159]

Nanotechnology Task Force

In 2006, then-acting FDA Commissioner Andrew von Eschenbach created an internal FDA Nanotechnology Task Force to "determin[e] regulatory approaches that encourage the continued development of innovative, safe and effective FDA-regulated products that use nanotechnology materials."[160] Neither FDA nor the task force adopted a definition of "nanotechnology."[161] The agency has stated that it "believes that the existing battery of pharmacotoxicity tests is probably adequate for most nanotechnology products that [it] will regulate."[162] In 2007, FDA declined to adopt labeling requirements for products containing nanomaterials, stating that:

> [b]ecause the current science does not support a finding that classes of products with nanoscale materials necessarily present greater safety concerns than classes of products without nanoscale materials, the [FDA] does not believe there is a basis for saying that, as a general matter, a product containing nanoscale materials must be labeled as such. Therefore, [FDA] is not recommending that the agency require such labeling at this time. Instead, [FDA] recommends ... the following action: Address on a case-by-case basis whether labeling must or may contain information on the use of nanoscale materials.[163]

Therefore, FDA has not promulgated specific regulations requiring products that contain nanomaterials to be labeled accordingly. The Nanotechnology Task Force indicated that regulatory decisionmaking "depends in part on having staff with expertise" in the appropriate areas and recommended that FDA build in-house expertise.[164] Also in 2007, the Nanotechnology Task Force recommended the agency coordinate with other federal agencies, the private sector, and other countries on research and other activities "to increase scientific understanding and facilitate assessment of data needs for regulated products" and undertake actions such as the development of guidance documents.[165]

Draft Guidance Regarding the Use of Nanomaterials in FDA-Regulated Products

On June 14, 2011, FDA issued draft guidance with recommendations for industry on "Considering Whether an FDA-Regulated Product Involves the Application of Nanotechnology," including the implications of using nanomaterials on the regulatory status of a product or the product's "safety, effectiveness, or public health impact."[166] The draft guidance is intended to assist industry and others to identify potential consideration for "regulatory status, safety, effectiveness, or public health impact" that may arise with the application of nanotechnology in all FDA-regulated products, including cosmetics.[167] The agency states that it "does not categorically judge all products containing nanomaterials or otherwise involving the application of nanotechnology as intrinsically benign or harmful."[168] However, FDA also notes that "evaluations of safety, effectiveness or public health impact of such products should consider the unique properties and behaviors that nanomaterials exhibit."[169]

On April 25, 2012, FDA issued draft guidance on the "Safety of Nanomaterials in Cosmetic Products."[170] This draft guidance provides a general framework for (1) assessing the safety of cosmetic products; (2) points to consider in assessing the safety of nanomaterials in cosmetic products, including a schema for characterizing the properties of nanomaterials and considerations for toxicology testing; and (3) a summary of FDA's recommendations. It notes that the use of nanomaterials "may alter the bioavailability of the cosmetic formulation," and that "traditional safety tests...may not be fully applicable."[171] FDA concludes that the inclusion of nanomaterials in an FDA-regulated product may affect the quality, safety, effectiveness, and/or public health impact of a product, and encourages manufacturers to meet with the FDA to discuss the "test methods and data needed to substantiate the product's safety, including short-term toxicity and long-term toxicity data as appropriate."[172]

Voluntary Cosmetic Registration Program

As noted above, FDA does not currently have the authority to mandate registration of cosmetic facilities, in contrast with the statutory registration requirements for establishments that produce other products regulated by the agency. However, since 1974, FDA, in cooperation with the cosmetic industry, has had a Voluntary Cosmetic Registration Program (VCRP) to facilitate registration of cosmetic establishments.[173] GAO has noted that "[r]egistration is important because it serves as the basis for determining where FDA will conduct its inspections."[174] FDA has also stated that VCRP information helps the Cosmetic Ingredient Review program (discussed previously) "in determining its priorities for ingredient safety review."[175]

Under VCRP, FDA encourages cosmetic establishments that manufacture or package cosmetic products to voluntarily register their facilities within 30 days of the start of their operations, regardless of whether their products enter interstate commerce.[176] FDA regulations request that foreign cosmetic product manufacturers voluntarily register with the agency if their products are exported for sale in the United States.[177] Cosmetic manufacturers and packagers also are encouraged to report the ingredients used in their product formulations.[178] FDA does not assess a fee for the voluntary registration of a cosmetic product establishment.[179]

Certain classes of establishments are exempt from FDA's voluntary registration request "because the [FDA] Commissioner has found that such registration is not justified."[180] These include beauty shops; cosmetologists; retailers; pharmacies; physicians; hospitals; clinics; public health agencies; persons who compound cosmetics at a location but do not otherwise manufacture or package cosmetics from that location; and persons who manufacture, prepare, compound, or process cosmetic products for activities such as teaching or research, but not for sale.[181]

Consumer safety organizations such as the Environmental Working Group have submitted comments to the FDA supporting the inclusion of "for professional use only" products in the voluntary registration scheme, particularly in light of issues with "Brazilian Blowout" products (see section """For Professional Use Only" Labeling" and the Appendix).[182] In its response to the comments, FDA disagreed with the inclusion of professional use products in the VCRP, as the VCRP does not apply to products not in commercial distribution.[183]

FDA also disagreed with the suggested audit of the cosmetics industry, which the consumer group proposed in order "to determine the current participation rate" in the VCRP and "to estimate how many ingredients and products FDA receives into the database compared to the total produced."[184] The agency focused on its lack of "statutory authority to make registration in the VCRP mandatory," as well as "the cost of completing such a project," calling the audit "not a wise use of Agency funds in the current economic environment."[185] Finally, FDA disagreed "at this time" with the Environmental Working Group's suggestion to create a certification program so that cosmetic companies could "indicate to consumers that they have participated in the VCRP," stating that the agency would need to research "how consumers would interpret such a certification claim," as well as how to enforce registration claims.[186]

REPORTING OF ADVERSE REACTIONS TO COSMETICS

FDA lacks the statutory authority to require cosmetic manufacturers to notify FDA of adverse events associated with their products and to require cosmetic companies to report information they receive from consumers and others regarding adverse events. Currently, the agency advises consumers to self-report "negative reaction[s] to a beauty, personal hygiene, [and] makeup products" to the FDA via the agency's safety information and adverse event reporting program— MedWatch[187]—or the consumer's local FDA complaint coordinator.[188] The agency is interested in hearing from consumers who "experience a rash, hair loss, infection, or other problem—even if they didn't follow product directions," as well as when products have bad smells or unusual colors and may be contaminated.[189] The agency may use adverse event reports by consumers to detect repeated problems with a product and potentially to take enforcement or other legal action.[190]

For example, adverse events that have been reported to FDA include reactions to henna/mehndi, certain shades of ink used for tattoos and permanent makeup, and keratin hair treatment products.[191] Temporary tattoos have been associated with reports of allergic reactions.[192] These products also have been subject to an import alert due to the lack of a required ingredient declaration on the label or the presence of colors not approved for use in cosmetics for the skin.[193]

Certain ink shades used for permanent makeup resulted in "more than 150 reports of adverse reactions in consumers."[194]

FDA has also received at least 33 adverse event reports, an additional seven reports of hair loss, and a number of inquiries concerning the safety of "Brazilian Blowouts" and similar "For Professional Use Only" hair treatment products, which may contain or release formaldehyde in the air when used by stylists to smooth hair, despite being labeled as "formaldehyde-free."[195] The Occupational Safety and Health Administration (OSHA), which regulates workers' exposure to formaldehyde and workplace safety, and state agencies that regulate hair salons have issued hazard alerts about these products.[196] This issue is discussed further in the **Appendix**.

In the absence of FDA requirements regarding adverse event reporting, the cosmetic industry has made efforts to self-regulate. In 2007, the industry trade association, the Personal Care Products Council (PCPC), created a Consumer Commitment Code that cosmetic product and ingredient manufacturers and marketers were "encouraged to acknowledge their support of" in writing.[197] One of the Code's principles is that "a company should notify the [FDA] of any known serious and unexpected adverse event as a result of the use of any of its cosmetic products marketed and used in the United States," where the terms "serious" and "unexpected" mean the same as FDA regulations defining serious and unexpected adverse events for drugs.[198] This Code is not a binding legal standard and cannot be enforced by FDA. The PCPC has stated that it "will not terminate the Council's membership for noncompliance," but would instead encourage compliance with the Code.[199]

OTHER CONCERNS WITH LABELING

Consumers may seek out particular cosmetics based on their labeling, such as cosmetics made with organic ingredients or without being tested on animals. However, FDA does not define certain terms used by manufacturers on their cosmetic products. Sections below on "organic" and "not tested on animals" claims address slight differences in how cosmetic products are marketed using certain claims and what consumers may believe such claims to mean. Additionally, not all cosmetic products are required to be labeled in the same manner, as the section below on products used by professionals discusses.

"Organic" Labeling Claims on Cosmetic Products

As with many statements made on cosmetic products, the terms "natural" and "organic" have no specific definition in the FFDCA, which may lead to consumer confusion.[200] While FDA has authority for labeling of cosmetics, the agency does not regulate the use of the term "organic"— rather, USDA regulates "organic" claims on cosmetic products.[201] Generally speaking, some cosmetics may be labeled as "natural" and "market[ed] ... as containing plant or mineral ingredients," while other cosmetic labels may include the claims that they are "organic" or made from "agricultural ingredients grown without pesticides."[202] Consumers seeking "natural" or "organic" cosmetics may have different expectations about the materials in a product marketed as natural or organic.

Consumers may perceive that products that are labeled as "natural" or "organic" have a health benefit.[203] However, FDA has noted that "many plants, regardless of whether they are organically grown, contain substances that may be toxic or allergenic."[204] Additionally, FDA has stated that "[c]onsumers should not necessarily assume that an 'organic' or 'natural' ingredient or product would possess greater inherent safety than another chemically identical version of the same ingredient."[205] Some natural ingredients may cause consumers to have adverse reactions, and FDA has stated that "[i]n fact, 'natural' ingredients may be harder to preserve against microbial contamination and growth than synthetic raw materials."[206]

In 2005, the USDA's National Organic Program (NOP), which oversees voluntary organic labeling of certified foods, determined that cosmetic products that meet the requirements established under the NOP regulations[207] are eligible for certification as "organic."[208] A cosmetic product "may be eligible to be certified under the NOP regulations" if the product "contains or is made up of agricultural ingredients, and can meet the USDA/NOP organic production, handling, processing and labeling standards."[209] The USDA has stated that the "organic" label is not meant to be an indicator of safety: "The National Organic Program is a marketing program, not a safety program."[210]

The NOP regulations provide four organic labeling categories: (1) 100% Organic—excluding water and salt, the product must be made of only organically produced ingredients and may use the USDA organic seal; (2) Organic—excluding water and salt, the product must be comprised of at least 95% organically produced ingredients and may use the USDA organic seal; (3) Made with Organic Ingredients—excluding water and salt, the product must contain at least 70% organic ingredients and the label may list three of the organic ingredients or food groups, such as herbs, but the product may not use the USDA organic seal; and (4) specific ingredients may be identified as organic if they are USDA-certified organic, but these products may not use the USDA organic seal or the term "organic."[211]

In 2009, the Certification, Accreditation, and Compliance Committee of the USDA's 15-member National Organics Standards Board made recommendations regarding "the problem of mislabeled organic personal care products."[212] The committee stated that the "USDA is responsible for product organic claims but is not currently enforcing this in the area of personal care products."[213] For example, some shampoos and conditioners state that they "use ingredients that are 100% Organic or are directly traceable to a natural source," but do not indicate who performs the organic certification or display the USDA Organic Seal.[214] As a result, the committee noted that "[c]onsumers are not assured that organic claims are consistently reviewed and applied" to personal care products.[215] The committee recommended amending the NOP regulations to include a definition of "personal care products" that is based on the definition of a "cosmetic" under the FFDCA, to clarify the use of the term "organic" in its application to personal care products, and to restrict the use of the USDA Organic Seal.[216] However, the recommendations of the committee have not yet been adopted by the National Organic Standards Board and "are not official USDA policy" at this time.[217]

In addition to the USDA's NOP, other entities have created their own standards programs for what constitutes "organic" in personal care products. For example, with input from industry stakeholders, the National Sanitation Foundation (NSF) International[218] and the American National Standards Institute (ANSI)[219] established a new nonfederal, voluntary standard, NSF/ANSI 305-2009e, for personal care products containing organic ingredients.[220] The standard allows a labeling claim of "contains organic ingredients" to be made for products with 70% or higher organic content, if the products comply with the standard's requirements, which include certification based on steps such as an application, on-site inspection, and technical review.[221] The standard requires manufacturers to list the exact percent of organic content.[222] The standard can be used for "rinse-off and leave-on personal care and cosmetic products, as well as oral care and personal hygiene products" if such products comply with "materials, processes, production criteria, and conditions" specified in the standard.[223] The major difference between the USDA NOP regulations and the NSF/ANSI standard is that the standard "allows for limited chemical processes that are typical for personal care products," which are "methods considered synthetic under the NOP."[224] According to NSF International, compliance with this standard may "provide a competitive advantage to those certified products" that contain organic ingredients.[225]

"Not Tested on Animals" Labeling

Many cosmetic products may contain ingredients or raw materials that have been tested on animals in the past, though no animal testing of the ingredients or product currently may be occurring.[226] While manufacturers may use "no animal testing" claims for their products, they still "may rely on raw material suppliers or contract laboratories to perform any animal testing necessary to substantiate product or ingredient safety."[227] It may be confusing for consumers attempting to distinguish cosmetic products with ingredients that have never been tested on animals from cosmetic products that may use or contract for the use of animal testing at some point in the product's path to commerce.[228] Some companies promote their products as not having been tested on animals, either because they contain all-natural ingredients or by labeling with such terms as "finished product not tested on animals," "no animal ingredients," or "cruelty free." FDA does not define or prescribe the use of these terms. In the absence of federal regulation on the use of such terms, animal rights groups have created programs where companies that self-certify that they are "cruelty free" may license the organization's logo for use on their products.[229]

"For Professional Use Only" Labeling

Certain information that is not required to appear in cosmetic product labeling may nonetheless be of interest to consumers and professionals who use and apply "for professional use only" cosmetic products. (The **Appendix** discusses the hazards potentially associated with one type of "for professional use only" product applied in keratin hair treatments, which are also known as Brazilian Blowouts.) This section provides general background on "for professional use only" cosmetic products. Cosmetics that are "consumer commodities" are required to list their ingredients, according to FDA regulations implementing the FPLA.[230] The ingredient listing requirement applies to products produced or distributed for retail sale and does not apply to "for professional use only" products used only by salons, if the salon does not also offer the product for purchase by its customers.[231] As a result, "cosmetologists and other professionals, as well as their clients, may not know what chemicals are in the cosmetics used in nonretail businesses, such as beauty salons."[232] However, if a cosmetic product were labeled "for professional use only" but sold at retail, the ingredients must be listed, or the cosmetic product will be considered to be misbranded.[233] Ingredients used in "for professional use only" cosmetic products are not included in the VCRP.[234] FDA does not define which cosmetic products are "For Professional Use Only." Cosmetic manufacturers and beauty supply companies that produce these products may limit distribution of such products to salons and salon professionals.[235] Despite manufacturer sale restrictions, some distributors have sold "for professional use only" products to retail stores, potentially in contravention of contracts or agreements between distributors and manufacturers regarding the sale of such products, as well as the misbranding prohibition of the FFDCA and related provisions in the FPLA.[236]

CONCLUSION

Although FDA's authorities over cosmetic products include some of those applicable to other FDA-regulated products, they are generally less comprehensive and exclude certain requirements imposed on other FDA-regulated products. The manner in which a cosmetic product could or should be regulated, however, is not always clear. FDA has issued regulations and procedures for cosmetics with which manufacturers voluntarily may choose to comply. Additionally, the cosmetic industry's trade association has established a cosmetic ingredient review program for cosmetic manufacturers with the purpose of determining which cosmetic ingredients are safe under certain conditions of use. Nevertheless, some questions remain as to whether the FDA's current oversight of cosmetic products and their ingredients is appropriate.

APPENDIX. KERATIN HAIR TREATMENTS, ALSO KNOWN AS "BRAZILIAN BLOWOUTS"

Background

Keratin hair treatment products reportedly smooth frizzy hair, straighten curly hair, and reduce blow drying and straightening times. Such treatments may also be known as "Brazilian Blowouts" after the name of one company's products commonly used for such treatments. The treatments typically cost several hundred dollars, depending on the length and texture of one's hair, and may last from six weeks to several months, depending on the type of treatment. Many brands of keratin hair treatment products have been found to contain free formaldehyde in solution (which tends to combine with water, forming methylene glycol), or other chemicals that convert into formaldehyde gas, whether or not they are labeled as "formaldehyde-free."[237] Formaldehyde and a related chemical, methylene glycol, are "known to induce a fixative action on proteins (e.g., keratin)," and therefore hair straightening solutions reportedly "maintain straightened hair by altering protein structures via amino acid crosslinking reactions, which form crosslinks between hair keratins and with added keratin from the formulation" of the hair product.[238]

Questions have been raised about the accuracy of ingredient statements and the adequacy of safety warnings on product labels for keratin hair treatment products containing formaldehyde.[239] Formaldehyde is a respiratory irritant and a known human carcinogen. The concern is that stylists who use such products, and consumers who are treated with them, may be exposed to harmful levels of formaldehyde without their informed consent, because many products are labeled "formaldehyde-free." As discussed below, investigations by the National Institute for Occupational Safety and Health (NIOSH), the Occupational Safety and Health Administration (OSHA), and Health Canada have indicated that even products labeled "formaldehyde-free" may contain levels of the chemical considered potentially unsafe.[240] While OSHA regulates workers' exposure to formaldehyde and worker and workplace safety, as discussed below, FDA regulates cosmetic products containing formaldehyde.[241]

Members of Congress have requested that FDA take enforcement actions against such keratin hair treatment products,[242] and FDA has issued a warning letter indicating certain Brazilian Blowout products are in violation of the FFDCA.[243] FDA is evaluating hair straightening and hair smoothing products for safety on an individual basis.[244] The manufacturer of Brazilian Blowout products has argued that testing by OSHA and "alternate reputable institutions" indicated that its products fall below OSHA safety standards.[245] OSHA has responded by asking the CEO of Brazilian Blowout to issue corrective statements to salon owners that "clearly stat[e] that OSHA air quality tests conducted ... have yielded results above acceptable OSHA limits."[246]

Formaldehyde

The Environmental Protection Agency (EPA) attempts to quantify the risk that an individual will suffer adverse health effects due to particular levels of exposure to a chemical. According to EPA, formaldehyde:

> can cause watery eyes, burning sensations in the eyes and throat, nausea, and difficulty in breathing in some humans exposed at elevated levels (above 0.1 parts per million). High concentrations may trigger attacks in people with asthma. There is evidence that some people can develop a sensitivity to formaldehyde. It has also been shown to cause cancer in animals and may cause cancer in humans. Health effects include eye, nose, and throat irritation; wheezing and coughing; fatigue; skin rash; severe allergic reactions. May cause cancer.[247]

The federal Agency for Toxic Substances and Disease Registry (ATSDR) concurs and adds that exposure may lead to:

> neurological effects, and increased risk of asthma and/or allergy ... in humans breathing 0.1 to 0.5 [parts formaldehyde per million parts of air (ppm)]. Eczema and changes in lung function have been observed at 0.6 to 1.9 ppm. Decreased body weight, gastrointestinal ulcers, and liver and kidney damage were observed in animals orally exposed to 50–100 mg/kg/day formaldehyde.[248]

The 12th Report on Carcinogens (ROC), issued by the U.S. Department of Health and Human Services' (HHS) National Toxicology Program (NTP) in June 2011 changed the classification of formaldehyde from "reasonably anticipated to be a human carcinogen" to "known to be a human carcinogen," based on its criterion that there is "sufficient evidence of carcinogenicity from studies in humans, which indicates a causal relationship between exposure to the agent, substance, or mixture, and human cancer."[249] This ROC listing does not necessarily mean that formaldehyde will cause an exposed individual to develop cancer; rather, it means that at some sufficient level of exposure to formaldehyde some humans will develop cancer.

The World Health Organization's International Agency for Research on Cancer (IARC) listed formaldehyde as a carcinogen in 2006.[250] OSHA recognizes the IARC list of carcinogens as well as the NTP ROC list for the purposes of its hazard communication standard, discussed below.

OSHA Formaldehyde Standards

Workers' exposure to formaldehyde in general industries as well as shipyard employment and construction is regulated at the federal level and is addressed in OSHA standards or equivalent regulations in OSHA-approved state plans.[251] OSHA has issued rules on formaldehyde exposure limits, protective equipment, and cancer warning labels for products that contain formaldehyde.[252] The agency also notes that "[s]hort-term exposure to formaldehyde can be fatal," and that "[l]ong-term exposure to low levels of formaldehyde may cause respiratory difficulty, eczema, and sensitization."[253]

OSHA's formaldehyde standard "applies to all occupational exposures to formaldehyde, i.e. from formaldehyde gas, its solutions, and materials that release formaldehyde."[254] OSHA's formaldehyde standard states that "[t]he permissible exposure limit (PEL) for formaldehyde in the workplace is 0.75 parts formaldehyde per million parts of air (0.75 ppm) measured as an 8-hour time-weighted average."[255] The standard also has short-term exposure limits of

2 ppm per 15- minute time period and sets a level at which "increased industrial hygiene monitoring and initiation of worker medical surveillance" is triggered.[256] OSHA notes that an "airborne concentration of formaldehyde above 0.1 ppm can cause irritation of the respiratory tract."[257]

Employers who have workplaces covered by the OSHA standard are required to monitor their employees' exposure to formaldehyde.[258] OSHA requires communication of formaldehyde's potential health hazards for "[f]ormaldehyde gas, all mixtures or solutions composed of greater than 0.1 percent formaldehyde, and materials capable of releasing formaldehyde into the air, under reasonably foreseeable conditions of use, at concentrations reaching or exceeding 0.1 ppm."[259] Employers are required to ensure that such products have hazard warning labels if they are "capable of releasing formaldehyde at levels of 0.1 ppm to 0.5 ppm," and if the products are "capable of releasing formaldehyde at levels above 0.5 ppm," the labels must contain additional information and the words "Potential Cancer Hazard."[260] Additionally, manufacturers and distributors of formaldehyde-containing products that meet the 0.1 percent level must "assure that material safety data sheets and updated information are provided to all employers purchasing such materials."[261] Based on a settlement with the California Attorney General, the website for the Brazilian Blowout products now contains a Material Safety Data Sheet for Brazilian Blowout Acai Professional Smoothing Solution, which indicates that the product is classified as a hazardous substance and warns about using proper ventilation.[262]

Adverse Event Reports

Hair salon stylists in Oregon first raised concerns about a hair smoothing product labeled "formaldehyde-free" when they began experiencing nosebleeds within a month of using the product and reportedly later developed chest pain and sore throats.[263] One stylist contacted the Oregon Health and Science University's Center for Research on Occupational and Environmental Toxicology, which conducted an investigation in 2010 with the Oregon Occupational Safety and Health Division.[264] Researchers found significant formaldehyde levels in 105 samples of hair smoothing treatments from 54 different salons.[265] Oregon's Occupational Safety and Health Division then issued alerts about the formaldehyde levels to over 21,000 state-licensed hair stylists.[266] Although products such as the Brazilian Blowout Acai Professional Smoothing Solution were labeled "formaldehyde-free," the tests found that the products had an average formaldehyde content of more than 8%.[267] Some products contained amounts of formaldehyde "well above what could legally be labeled as 'formaldehyde-free.'"[268]

Oregon's Occupational Safety and Health Division received reports of adverse events from stylists across the United States after its alert, which included "burning of eyes and throat, watering of eyes, dry mouth, loss of smell, headache and a feeling of 'grogginess,' malaise, shortness of breath and breathing problems, a diagnosis of epiglottitis attributed by the stylist to their use of the product, fingertip numbness, and dermatitis," as well as reports of hair loss.[269] FDA has received reports from state and local groups of "eye irritation, breathing problems, and headaches," as well as adverse event reports from "hair stylists, their customers, and individual users" of similar symptoms, plus fainting, bronchitis, inhalation pneumonitis, and vomiting.[270] Similarly, Health Canada reportedly received adverse reaction reports for hair products with formaldehyde from 50-60 individuals, which included "burning eyes, nose, throat and breathing difficulties, with one report of hair loss," as well as reports of "headache, arthritis, dizziness, epistaxis [nosebleeds], swollen glands, and numb tongue."[271]

NIOSH and OSHA Investigations

In December 2010, NIOSH conducted a health hazard evaluation of the Brazilian Blowout Acai Professional Smoothing Solution, as used by one hair stylist employee on another hair stylist in a salon.[272] The evaluation indicated that the solution's concentration of formaldehyde (greater than 0.1%) was enough to merit the "hazard communication requirements of the OSHA formaldehyde standard."[273] OSHA has conducted its own investigations of keratin treatment products.[274] OSHA's investigations "found formaldehyde in the air when stylists used hair smoothing products," even though not all of the products had "formaldehyde listed on their labels or in material safety data sheets as required by law."[275] OSHA air tests of one product labeled as "formaldehyde-free" exceeded OSHA's limits on formaldehyde.[276] OSHA has issued at least one citation to an employer after air sampling found that salon workers "were exposed to formaldehyde levels that exceeded OSHA's 15-minute short term exposure limit."[277]

OSHA issued a Hazard Alert to hair salons indicating the hazards associated with the use of hair smoothing treatment products and the responsibilities of salons that use these products under the federal

Occupational Safety and Health Act.[278] California and several other states have issued similar notices.[279] In August 2011, the CEO of Brazilian Blowout sent a letter to salon owners indicating that "all OSHA and independent air-quality tests conducted on the Brazilian Blowout Professional Smoothing Solution ... have yielded results well-below even the most stringent of OSHA standards."[280] In September 2011, OSHA issued a letter to the Brazilian Blowout CEO informing him that OSHA disagreed with his remarks and requesting that he immediately take corrective actions such as sending a correction or retraction to his letter to salon owners, "clearly stating that OSHA air quality tests conducted ... have yielded results above acceptable OSHA limits."[281]

Actions by Other Countries

In 2011, Health Canada issued an advisory naming eleven keratin or similar smoothing hair treatment products with levels of formaldehyde ranging from 0.35% to 8.4%, which exceed the level of 0.2% at which it is "permitted as a preservative" in Canada.[282] Therefore, "hair smoothing products with formaldehyde levels" above 0.2% are banned from being sold in Canada.[283] This 0.2% level for formaldehyde and its equivalents is also the upper limit recommended by the Cosmetic Ingredient Review panel.[284] Authorities in France and Germany have warned against the use of hair smoothing products with high concentrations of formaldehyde, and both France and Ireland took steps to remove products from the market.[285]

Cosmetic Ingredient Review Analysis

The Cosmetic Ingredient Review (CIR) recently re-evaluated the safety of formaldehyde and addressed the safety of methylene glycol, a compound formed when formaldehyde is combined with water, in cosmetic products.[286] As mentioned in the body of this chapter, under the CIR program, expert panels analyze information on the safety of ingredients used in cosmetic products. In October 2011, CIR issued a final amended report on formaldehyde and methylene glycol that stated that "[n]ot surprisingly, formaldehyde is an irritant at low concentration, especially to the eyes and the respiratory tract. Formaldehyde exposure can result in a sensitization reaction."[287] CIR stated that its panel "continues to believe that formaldehyde gas can produce [nasopharyngeal] cancers at high doses."[288]

CIR's expert panel "was concerned" with adverse event reports, which it noted were "consistent with measured air levels of formaldehyde in salons" using hair straightening products and indicated that not all ventilation controls were effective in allowing for safe use.[289] CIR cited the Oregon Occupational Safety and Health Division's workplace survey of ventilation efforts that ranged from "a building HVAC system, propping the business's doors open, or operating ceiling fans."[290]

The CIR panel concluded that "[i]n the present practices of use and concentration (on the order of 10% formaldehyde/methylene glycol, blow drying and heating up to 450°F with a flat iron, inadequate ventilation, resulting in many reports of adverse effects), hair smoothing products containing formaldehyde and methylene glycol are unsafe."[291] However, CIR found that formaldehyde and methylene glycol "are safe for use in cosmetics when formulated to ensure use at the minimal effective concentration, but in no case should the formalin [formaldehyde and water solution] concentration exceed 0.2%."[292] As an example, the panel discussed the use and concentration of formaldehyde and methylene glycol in nail hardening products.[293]

An Assessment of FDA's Authorities

Some Members of Congress and the chief scientist of an industry trade association have asked FDA to take action on keratin hair treatment products.[294] This section discusses FDA's existing authorities and potential actions that the agency could take with regard to such cosmetic products, as well as the warning letter that FDA has issued to the CEO of Brazilian Blowout and actions by the California Attorney General. FDA does not have authority to regulate "the operation of salons or the practice of cosmetology."[295]

FDA does not ban formaldehyde or methylene glycol in cosmetic products.[296] According to the CIR, FDA's voluntary cosmetic registration program contained 77 uses of formaldehyde and formaldehyde solution (formalin).[297] FDA has stated that the safety of formaldehyde "as a cosmetic ingredient depends on a variety of factors, such as its concentration in the final product and how the final product is used."[298] FDA could issue a rule prohibiting or restricting the use of formaldehyde and formaldehyde solutions in cosmetic products if the agency concluded that such substances were poisonous or deleterious.[299] If such ingredients were deemed deleterious, their inclusion in a cosmetic product would render the product adulterated under the FFDCA.[300] It is a prohibited act to introduce an adulterated product into interstate commerce under the FFDCA and such an action may subject an individual or company to criminal penalties.[301]

FDA does not require a warning label on cosmetic products containing formaldehyde, formalin, methlyene glycol, or related chemicals. However, FDA is authorized to conduct rulemaking to require a warning statement on cosmetic products with such ingredients. FDA regulations provide that "[t]he label of a cosmetic product shall bear a warning statement whenever necessary or appropriate to prevent a health hazard that may be associated with the product."[302]

FDA may take enforcement actions against adulterated or misbranded cosmetic products, such as cosmetic products with misleading labels. Labeling must be deemed to be misleading if it does not reveal material facts "in light of other representations made or suggested by statement, [or] word."[303] FDA has indicated that the omission of material facts on the labeling of keratin hair treatment products—i.e. labeling these products "formaldehyde-free" when they in fact contain formaldehyde—could make such products misbranded under the FFDCA.[304]

In August 2011, FDA issued a warning letter to the CEO of Brazilian Blowout, noting that the product was both adulterated and misbranded under the FFDCA.[305] FDA asserted that the product was adulterated because the cosmetic "bears or contains a deleterious substance [methylene glycol] that may render it injurious to users under the conditions of use prescribed in your labeling."[306] Additionally, FDA stated that the product was misbranded because "its label and labeling (including instructions for use) makes misleading statements regarding the product's ingredients and fails to reveal material facts with respect to consequences that may result from the use of the product."[307] FDA advised the CEO to take corrective actions or face potential enforcement actions, including seizures and injunctions, and emphasized that manufacturers have a duty to ensure the products they market are safe and in compliance with FDA requirements.[308]

Depending on how "formaldehyde free" hair keratin products have been advertised, the Federal Trade Commission also may be authorized to initiate an action for deceptive advertising.[309] Action by state attorneys general may also be possible. The California Attorney General's office filed a lawsuit against one company for labeling violations, deceptive advertising, and violations of state cosmetics and toxics acts.[310] The lawsuit resulted in a settlement with the manufacturer requiring a "CAUTION" warning on two of its products (including a California Proposition 65 cancer warning); the production of a Material Safety Data Sheet and its posting on the company's website; the end of deceptive advertising, including modifications to the company's website; retesting of products at approved laboratories; reporting to the California Department of Public Health Safe Cosmetics Program; the disclosure of refund policies; proof of professional licensing before sale of professional use only products; civil penalties; and attorneys fees.[311]

As discussed earlier in this chapter, FDA does not have the authority to require premarket approval or premarket review of cosmetic ingredients or cosmetic products, except for color additives.[312] Additionally, FDA cannot mandate that a company recall a product that may violate the FFDCA or FPLA, but the agency can request that a manufacturer voluntarily recall a cosmetic product.[313] Nor does the agency have the authority to mandate adverse event reports for interactions that consumers experience from the use of a company's products. However, as indicated earlier, FDA has encouraged consumer reporting of adverse events associated with cosmetics. The agency's website discusses complaints regarding the use of Brazilian Blowout and other hair smoothing products and urges consumers and salon professionals to report adverse events to FDA.[314]

Finally, FDA cannot require professional use cosmetic products, such as Brazilian Blowout, to list their ingredients if they are not "consumer commodities"—products produced or distributed for retail sale—under the FPLA.[315] However, if a cosmetic product was labeled "for professional use only" but sold at retail, the ingredients must be listed, or the cosmetic will be considered to be misbranded.[316] FDA stated in November 2010 that it was "investigating whether or not Brazilian Blowout is marketed directly to consumers. If so, failure to comply with the ingredient declaration requirement would constitute misbranding."[317]

End Notes

[1] First Research, Industry Profile, Cosmetics, Beauty Supply, and Perfume Stores, May 23, 2011.

[2] Personal Care Products Council, About Us, http://www.personalcarecouncil.org/about-us/about-personal-careproducts-council.

[3] U.S. Census Bureau, 2007 Economic Census, Sector 44: Retail Trade: Industry Series: Preliminary Summary Statistics for the United States, Cosmetics, Beauty Supplies, and Perfume Stores, Sept. 29, 2009.

[4] First Research, *supra* note 1.

[5] First Research, Industry Profile, Personal Care Products Manufacturing, May 16, 2011.

[6] Loran Braverman, CFA, Standard & Poors, NetAdvantage, Sub-Industry Review: Personal Products, http://www.netadvantage.standardandpoors.com/NASApp/NetAdvantage/showSubIndustryReview.do?subindcode= 30302010; First Research, *supra* note 5.

[7] First Research, *supra* note 5.

[8] Department of Health and Human Services (HHS), Fiscal Year 2013, *Food and Drug Administration, Justification of Estimates for Appropriations Committees*, http://www.fda.gov/downloads/AboutFDA/ReportsManualsForms/Reports/BudgetReports/default.htm, p. 103.

[9] 21 U.S.C. §§ 361, 362; Federal Food, Drug, and Cosmetic Act (FFDCA) §§ 601, 602.

[10] Under the Federal Trade Commission Act, "[i]t shall be unlawful for any person ... to disseminate, or cause to be disseminated, any false advertisement—(1) By United States mails, or in or having an effect upon commerce, by any means, for the purpose of inducing, or which is likely to induce, directly or indirectly the purchase of ... cosmetics; or (2) By any means, for the purpose of inducing, or which is likely to induce, directly or indirectly, the purchase in or having an effect upon commerce, of ... cosmetics." 15 U.S.C. § 52. Additionally, cosmetics are explicitly excluded from the definition of "consumer product" in the Consumer Product Safety Act, which is enforced by the Consumer Product Safety Commission. 15 U.S.C. § 2052(a)(5)(H).

[11] 21 U.S.C. § 359; FFDCA § 509.

[12] Examples of cosmetics "that may be introduced into the body are limited, but include mouthwashes, breath fresheners, and vaginal douches." John E. Bailey, *Organization and Priorities of FDA's Office of Cosmetics and Colors*, Cosmetic Regulation in a Competitive Environment, Norman F. Estrin & James M. Akerson, eds., p. 217, 2000.

[13] 21 C.F.R. § 720.4(c)(12).

[14] 21 U.S.C. § 321(i); FFDCA § 201(i).

[15] The FDA has defined soap in its regulations as applying only to articles for which "(1) [t]he bulk of the nonvolatile matter in the product consists of an alkali salt of fatty acids and the detergent properties of the article are due to the alkali-fatty acid compounds; and (2) [t]he product is labeled, sold, and represented only as soap." 21 C.F.R. § 701.20(a). A product intended not only for cleansing but also for other cosmetic uses such as beautifying, moisturizing, or deodorizing would be regulated by FDA as a cosmetic. A soap-like product may also be a drug, if it is intended to cure, treat, or prevent disease or to affect the structure or any function of the human body. 21 U.S.C. § 321(i)(2); FFDCA § 201(i)(2).

[16] 21 U.S.C. § 361(a); FFDCA § 601(a).

[17] 21 U.S.C. § 321(g); *see* Amity Hartman, *FDA's Minimal Regulation of Cosmetics and the Daring Claims of Cosmetic Companies that Cause Consumers Economic Harm*, 36 W. ST. L. REV. 53, 58 (2008)(noting that manufacturer intentions affect the classification of a products). The intended use of a product is displayed by several factors including claims stated on the product labeling, in advertising, or other promotional materials; consumer perception and the products reputation; and, ingredients that may cause the product to be considered a drug by industry standards or public perception. *See* FDA, Is It a Cosmetic, a Drug, or Both (Or Is It Soap?), July 8, 2002, http://www.fda.gov/Cosmetics/GuidanceComplianceRegulatoryInformation/ucm074201.htm.

[18] 21 U.S.C. § 355; FFDCA § 505. A new drug application (NDA) is the process through which drug sponsors propose that FDA approve a new pharmaceutical for sale and marketing in

the United States. Among other considerations, the agency approves a NDA after examining reports and investigations that demonstrate the drug's safety and effectiveness.

[19] 21 C.F.R. Part 350; 21 C.F.R. §§ 358.701-760.

[20] A monograph is a set of rules promulgated by the FDA for a number of OTC drug categories. These OTC drug monographs may state the types of active ingredients, including a list of specific active ingredients, indications, usage instructions, warnings and other labeling requirements for a given category of OTC drugs.

[21] 21 C.F.R. § 201.66(c).

[22] 21 U.S.C. § 355(a)(2)(B); FFDCA § 501(a)(2)(B).

[23] For information on some FDA requirements related to drugs, see FDA, Drug Application and Approval Process – "Questions and Answers," http://www.fda.gov/AboutFDA/Centers Offices/CDER/ucm197608.htm.

[24] "Cosmeceuticals combine cosmetics and pharmaceutical benefits, and may contain patented ingredients or have dermatologist endorsements. Cosmeceutical sales are expected to grow faster than the overall cosmetics and toiletries market, according to Scientia Advisors." First Research, Industry Profile, Cosmetics, Beauty Supply, and Perfume Stores, May 23, 2011.

[25] 21 C.F.R. § 70.3(a), (f) (setting forth the required designations of ingredients for the labeling of cosmetic products).

[26] 58 Fed. Reg. 28194, 28204 (May 12, 1993) "When an ingredient can be used for either drug or cosmetic purposes, its regulatory status as a drug or cosmetic, or both, is determined by objective evidence of the distributor's intent."

[27] 21 C.F.R. § 700.35 "A product that includes the term 'sunscreen' in its labeling ... comes within the definition of a drug. ... [T]he use of the term 'sunscreen' or similar sun protection terminology in a product's labeling generally causes the product to be subject to regulation as a drug."

[28] 21 U.S.C. § 331(a)-(c); FFDCA § 301(a)-(c).

[29] 21 U.S.C. § 361; FFDCA § 601.

[30] 21 U.S.C. § 362; FFDCA § 602. In addition to the FFDCA, the Fair Packaging Act and Labeling Act (FPLA) requires cosmetic labels to comply with specific guidelines. If cosmetics are found to be in violation of the FPLA statutory or regulatory provisions, they are considered misbranded for the purposes of the FFDCA. 15 U.S.C. § 1456(a); *see also* U.S. Food and Drug Administration, Key Legal Concepts: "Interstate Commerce," "Adulterated," and "Misbranded," http://www.fda.gov/Cosmetics/GuidanceCompliance RegulatoryInformation/ucm074248.htm.

[31] S. Comm. on Commerce, S. Rep. No. 91, 75th Cong., p. 5, 1937.

[32] Peter Barton Hutt, *A History of Government Regulation of Adulteration and Misbranding of Cosmetics*, *in* Cosmetic Regulation in a Competitive Environment, Norman F. Estrin & James M. Akerson eds. 2000.

[33] Ibid. at 2.

[34] Hutt, *supra* note 32, p. 5-6; Jacqueline A. Greff, *Regulation of Cosmetics That are Also Drugs*, 51 Food & Drug L. J. 243, 244 (1996).

[35] Hutt, *supra* note 32, p. 2-3, 6.

[36] 15 U.S.C. § 1451 *et seq.*

[37] 15 U.S.C. § 1459(a).

[38] 15 U.S.C. § 1452(b).

[39] Personal Care Products Council, A Dynamic Industry at Work: 2008 Annual Report, p. 5, http://www.personalcarecouncil.org/sites/default/files/2008CouncilAnnualReport.pdf; The

Campaign for Safe Cosmetics, State Legislation, http://safecosmetics.org/article.php?id=345.

[40] 21 U.S.C. §§ 331-334; FFDCA §§ 301-04.

[41] 21 U.S.C. § 381; FFDCA § 801.

[42] 21 U.S.C. § 374(a); FFDCA § 704(a).

[43] FDA, Inspection of Cosmetics: An Overview, http://www.fda.gov/Cosmetics/ GuidanceComplianceRegulatoryInformation/ComplianceEnforcement/ucm136455.htm.

[44] Ibid.

[45] Ibid.; FFDCA § 704(c).

[46] 21 C.F.R. Parts 710, 720; FDA, Bad Reaction to Cosmetics? Tell FDA, http://www.fda.gov/ForConsumers/ ConsumerUpdates/ucm241820.htm.

[47] 21 U.S.C. § 350d (food); 21 U.S.C. § 360 (drugs and devices); 21 U.S.C. § 387e (tobacco).

[48] FDA Authority Over Cosmetics, http://www.fda.gov/Cosmetics/GuidanceComplianceRegulatoryInformation/ ucm074162.htm; Donald R. Johnson, *Not in my Makeup: The Need for Enhanced Premarket Regulatory Authority Over Cosmetics in Light of Increased Usage of Engineered Nanoparticles*, 26 J. Contemp. Health L. & Policy 82, 114, 2009.

[49] 21 C.F.R. § 720.4.

[50] 21 C.F.R. Part 7, Subpart C.

[51] FFDCA §§ 301-04.

[52] The FDA's authority over cosmetic products is based primarily on the FFDCA provisions on cosmetics, color additives, and drugs. The agency also has authority under the FPLA for labeling requirements. Other agencies may use their own authorities to regulate certain aspects of cosmetic products, e.g., the Federal Trade Commission regulates the advertising of cosmetics.

[53] 21 C.F.R. Parts 210 and 211; FDA, Good Manufacturing Practice (GMP) Guidelines/Inspection Checklist, http://www.fda.gov/Cosmetics/GuidanceComplianceRegulatoryInformation/GoodManufacturingPracticeGMPGuidelinesInspection Checklist/default.htm. In 1977, the FDA issued a "notice of intent to propose regulations" for the "preservation of cosmetics coming in contact with the eye," in which the FDA indicated it "expects to promulgate all-inclusive regulations delineating good manufacturing practice for cosmetics at some point, and [the FDA Commissioner] intend[ed] to propose regulations regarding microbial preservation of cosmetics coming in contact with the eye as a first step. 42 Fed. Reg. 54837, 54837, Oct. 11, 1977. The industry trade association—Cosmetic, Toiletry and Fragrance Association, now the Personal Care Products Council—reportedly "filed a petition describing the industry's preferred cosmetic GMPs," which were reportedly included by FDA for a time into the agency's *Investigative Operations Manual*. Greff, *supra* note 31, p. 246.

[54] FDA, Inspection of Cosmetics: An Overview, http://www.fda.gov/Cosmetics/ GuidanceComplianceRegulatoryInformation/ComplianceEnforcement/ucm136455.htm.

[55] 21 U.S.C. § 379e; 21 U.S.C. § 359. Premarket approval for color additives was established in 1960 with the Color Additive Amendments of 1960. P.L. 86-618. A color additive is basically defined as a substance that, when added or applied to a cosmetic or the body, is capable of imparting coloring. Examples of cosmetics with color additives include lipstick, blush, and eye makeup. 21 U.S.C. § 321(t).

[56] 21 U.S.C. § 355; FFDCA § 505.

[57] 21 U.S.C. § 331(a)-(c); FFDCA § 301(a)-(c).

[58] 21 U.S.C. § 361; FFDCA § 601; 21 C.F.R. § 740.18. "The coal tar hair dye exemption allows coal tar hair dyes, not intended for use on eyelashes or eyebrows, to be marketed to

consumers, even if they have been found to be injurious to the user under conditions of use." Bailey, *supra* note 12, at 220. The label for coal tar hair dye products must contain the statutorily-required caution statement in order to not be considered to be adulterated, as well as "adequate directions for conducting such preliminary testing," which are not specified by the FDA, but rather have been set as a self-evaluation patch test with a wait time of 48 hours by the industry-established Cosmetic Ingredient Review (CIR). p. 219-20. If the coal tar hair dye product does not contain that information, the coal tar dye is "subject to regulation as a cosmetic coal additive and must be approved by FDA and listed in the CFR before marketing." p. 220. In 1952, a congressional committee report recommending the elimination of the coal tar hair dye exemption. Hutt, *supra* note 32, p. 25. The Government Accountability Office (GAO) also issued a report in the late 1970s recommending the elimination of this exemption. Ibid. p. 27.

[59] 21 C.F.R. § 700.19—Use of methylene chloride as an ingredient of cosmetic products.

[60] Henna is "a coloring made from a plant" that is directly applied to the skin "in the body-decorating process known as mehndi." FDA, Temporary Tattoos & Henna/Mehndi, http://www.fda.gov/Cosmetics/ProductandIngredientSafety/ProductInformation/ucm108569.htm.

[61] FDA Import Alert 53-19, Oct. 2, 2009.

[62] Ibid.

[63] 15 U.S.C. §§ 1451 *et seq.*

[64] 15 U.S.C. § 1456(a). However, while the FDA may take enforcement action against consumer commodity products that are considered misbranded because they do not conform to FPLA provisions, the penalty provisions of the FFDCA that could be sought for products deemed misbranded under the FFDCA do not apply to products deemed to be misbranded because they violate the FPLA's provision on unfair and deceptive packaging and labeling. That FPLA provision makes it unlawful for persons engaged in packing or labeling consumer commodities to distribute, or cause to be distributed, a consumer commodity in a package or with a label that does not meet the FPLA provisions. 15 U.S.C. § 1452(a); 15 U.S.C. § 1456(a).

[65] 21 C.F.R. § 701.11 (identity labeling); 21 C.F.R. § 701.12 (name and place of business or manufacturer, packer, or distributor); 21 C.F.R. § 701.13 (declaration of net quantity of contents); 21 C.F.R. 701.3 and 21 C.F.R. § 21.66 (designation of ingredients, including active drug ingredients if the cosmetic product is also an over-the-counter drug product); 21 C.F.R. § 1.21 (failure to reveal material facts on labeling); 21 C.F.R. Parts 700 and 740 (warning language or requirements for certain cosmetic products).

[66] 21 U.S.C. § 362; FFDCA § 602. FDA regulations provide that "[t]he labeling of a cosmetic which contains two or more ingredients may be misleading by reason ... of the designation of such cosmetic in such labeling by a name which includes or suggests the name of one or more but not all such ingredients, even though the names of all such ingredients are stated elsewhere in the labeling." 21 C.F.R. § 701.1(b). FDA regulations also provide that "[a]ny representation in labeling or advertising that creates an impression of official approval because of [the filing of Form FDA 2512, Cosmetic Product Ingredient Statement] will be considered misleading." 21 C.F.R. § 720.9.

[67] The principal display panel is "that part of a label that is most likely to be displayed, presented, shown, or examined under normal and customary conditions of display for retail sale." 15 U.S.C. § 1459(f); 21 C.F.R. § 701.10.

[68] 15 U.S.C. § 1453(a)(3); 21 C.F.R. § 701.2—Form of stating labeling requirements.

[69] 15 U.S.C. § 1454(c)(3); 21 C.F.R. § 701.3(a). However, the FDA's regulation does "not require the declaration of incidental ingredients that are present in a cosmetic at insignificant levels and that have no technical or functional effect in the cosmetic" such as processing aids. 21 C.F.R. § 701.3(l).

[70] Bailey, *supra* note 12, p. 218.

[71] Ibid.

[72] 21 U.S.C. §§ 331-334; FFDCA §§ 301-04.

[73] Nicole Abramowitz, *The Dangers of Chasing Youth: Regulating the Use of Nanoparticles in Anti-Aging Products*, 2008 U Ill. J.L. Tech. & Policy 199, p. 208-09, Spring 2008.

[74] 21 U.S.C. § 350l.

[75] FFDCA § 412(f).

[76] FFDCA § 518(e).

[77] 42 U.S.C. § 264; 21 C.F.R. § 1271.440.

[78] FFDCA § 908(c).

[79] 21 C.F.R. § 7.40(b); FDA Regulatory Procedures Manual, Ch. 7: Recall Procedures, http://www.fda.gov/downloads/ ICECI/ComplianceManuals/RegulatoryProceduresManual/UCM074312.pdf.

[80] 21 C.F.R. § 7.3(g); 21 C.F.R. § 7.40(a).

[81] 21 C.F.R. § 7.45.

[82] 21 C.F.R. § 7.53.

[83] 21 C.F.R. § 7.41.

[84] 21 C.F.R. §§ 7.42(b)(2), 7.50.

[85] 21 C.F.R. §§ 7.53, 7.55.

[86] FFDCA § 721; FDA, FDA Authority Over Cosmetics, http://www.fda.gov/Cosmetics/ GuidanceComplianceRegulatoryInformation/ucm074162.htm.

[87] FDA, FDA Authority Over Cosmetics, *supra* note 86. The FDA has said that "the safety of a product can be adequately substantiated through (a) reliance on already available toxicological test data on individual ingredients and on product formulations that are similar in composition to the particular cosmetic, and (b) performance of any additional toxicological and other tests that are appropriate in light of such existing data and information. Although satisfactory toxicological data may exist for each ingredient of a cosmetic, it will still be necessary to conduct some toxicological testing with the complete formulation to assure adequately the safety of the finished cosmetic." FDA, Cosmetic Products: Warning Statements/Package Labels, 40 Fed. Reg. 8912, 8916, Mar. 3, 1975; FDA, Cosmetics, Product Testing, http://www.fda.gov/Cosmetics/ProductandIngredient Safety/ ProductTesting/default.htm.

[88] 21 C.F.R. § 740.10.

[89] Bailey, *supra* note 12, p. 218, stating that "[n]o product has ever been encountered in retail commerce that bears the warning statement specified in 21 CFR 740.10."

[90] *The Food and Drug Administration's Regulation of Cosmetics*: *Before the Subcomm. on Oversight and Investigations of the House Comm. on Interstate and Foreign Commerce*, p. 4, Feb. 3, 1978, statement of Gregory J. Ahart, Director, Human Resources Division, GAO.

[91] 54 Fed. Reg. 27328, 27340, June 29, 1989; 21 C.F.R. § 700.19(b), "Any cosmetic product that contains methylene chloride as an ingredient is deemed adulterated and is subject to regulatory action under sections 301 and 601(a) of the [FFDCA]."

[92] For example, acetyl ethyl tetramethyl tetralin (AETT) was "voluntarily discontinued" by the fragrance industry in 1978 after it "was found to cause serious neurotoxic disorders and discoloration of internal organs" in a 1977 toxicity study of rats. FDA, Cosmetic Product

Manufacturers (2/95), Guide to Inspections of Cosmetic Product Manufacturers, http://www.fda.gov/ICECI/Inspections/InspectionGuides/ucm074952.htm.

[93] 21 C.F.R. § 740.10.

[94] FDA, Cosmetics and U.S. Law, http://www.fda.gov/Cosmetics/International Activities/ CosmeticsU.S.Law/ default.htm.

[95] 21 C.F.R. § 740.10; FDA, Cosmetics Q&A: Animal Testing, http://www.fda.gov/Cosmetics/ ResourcesForYou/ Consumers/CosmeticsQA/ucm167216.htm.

[96] Helen Northroot, *Substantiating the Safety of Cosmetic and Toiletry Products*, Cosmetic Regulation in a Competitive Environment, Norman Estrin & James Akerson, eds., 2000.

[97] FDA, Animal Testing, http://www.fda.gov/Cosmetics/ProductandIngredientSafety/ ProductTesting/ucm072268.htm.

[98] For example, The Humane Society of the United States, People for the Ethical Treatment of Animals (PETA), and the American Anti-Vivisection Society have campaigns against animal testing for cosmetics. Humane Society, "'Be Cruelty Free' Campaign Launches to End Cosmetics Testing on Animals," http://www.humanesociety.org/news/ press_releases/2012/04/be_cruelty_free_campaign?042312.html; PETA, Cosmetics and Household-Product Animal Testing, http://www.peta.org/issues/animals-used-for-experimentation/cosmetic-household-products-animaltesting.aspx; and, American Anti-Vivisection Society, Who We Are, http://www.aavs.org/site/c.bkLTKfOSLhK6E/ b.6452345/k.24B3/Who_We_Are.htm.

[99] FDA, Animal Testing, *supra* note 97; P.L. 89-544. (codified as amended at 7 U.S.C. § 2131 *et seq.*).

[100] FDA, Animal Testing, *supra* note 97.

[101] CIR, Cosmetic Ingredient Review Procedures, Oct. 2010, p. 4, http://www.cir-safety.org/pdf1.pdf.

[102] The Voluntary Cosmetic Registration Program is discussed in greater detail *infra*.

[103] CIR, Cosmetic Ingredient Review Procedures, Oct. 2010, p. 11, http://www.cir-safety.org/pdf1.pdf.

[104] CIR, General Information, How Does CIR Work?, http://www.cir-safety.org/info.shtml.

[105] Gary L. Yingling and Suzan Onel, *Cosmetic Regulation Revisited*, Fundamentals of Law and Regulation, Vol. I, p. 333, Robert P. Brady et al., eds.1997.

[106] CIR, Cosmetic ingredients found safe as used (through February 2012), available at http://www.cir-safety.org/ supplementaldoc/safe-used.

[107] CIR, Cosmetic ingredients found safe, with qualifications (through February 2012), http://www.cir-safety.org/ supplementaldoc/safe-qualifications.

[108] CIR, Cosmetic ingredients with insufficient data to support safety (through February 2012), http://www.cirsafety.org/cir-findings.

[109] CIR, Ingredients found unsafe for use in cosmetics (through February 2012), http://www.cir-safety.org/ supplementaldoc/unsafe-ingredients.

[110] RIFM, a nonprofit corporation, works in part to "encourage uniform safety standards related to the use of fragrance ingredients." The corporation reportedly has the world's largest database of flavor and fragrance materials, with more than 5,000 materials, RIFM, About Us, http://www.rifm.org/about.php.

[111] The Campaign for Safe Cosmetics, http://www.safecosmetics.org/.

[112] "Cosmetics containing as ingredients amines and amino derivatives ... may form nitrosamines, if they also contain an ingredient which acts as a nitrosating agent ... Many nitrosamines have been determined to cause cancer in laboratory animals. They have also been shown to penetrate the skin." FDA expressed its concern about the contamination of cosmetics with

nitrosamines in a Federal Register notice dated April 10, 1979, which stated that cosmetics containing nitrosamines may be considered adulterated and subject to enforcement action. FDA, Cosmetic Product Manufacturers: Guide to Inspections of Cosmetic Product Manufacturers, http://www.fda.gov/ICECI/ Inspections/InspectionGuides/ucm074952.htm.

[113] "FDA has not set limits for contaminants, such as lead, in cosmetics. However, FDA does set specifications for impurities, such as lead, for color additives used in cosmetics." FDA, Lipstick and Lead: Questions and Answers, http://www.fda.gov/Cosmetics/Productand IngredientSafety/ProductInformation/ucm137224.htm.

[114] FDA, Parabens, http://www.fda.gov/Cosmetics/ProductandIngredientSafety/Selected CosmeticIngredients/ ucm128042.htm. "Companies use parabens to extend the shelf life of products and prevent growth of bacteria and fungi in, for instance, face cream. ... [S]ome think that parabens may be linked to breast cancer and fertility issues." Alene Dawson, *'Paraben-free': Should You Care?,* LA Times, May 8, 2011, latimes.com/features/image/la-ig-beautyparabens-20110508,0,1400441.story.

[115] Phthalates are "a group of chemicals used in hundreds of products, such as ... nail polish, hair sprays, soaps, and shampoos." FDA, Phtalates and Cosmetic Products, http://www.fda.gov/Cosmetics/ProductandIngredientSafety/ SelectedCosmeticIngredients/ucm128250.htm. Also CRS Report RL34572, *Phthalates in Plastics and Possible Human Health Effects*, by Linda-Jo Schierow and Margaret Mikyung Lee.

[116] "Hydroquinone is a skin bleaching ingredient used to lighten areas of darkened skin." FDA, Hydroquinone Studies Under The National Toxicology Program (NTP), http://www.fda.gov/AboutFDA/CentersOffices/CDER/ ucm203112.htm.

[117] "The compound 1,4-dioxane is a contaminant that may be present in extremely small amounts in some cosmetics. It forms as a byproduct during the manufacturing process of certain cosmetic ingredients. ... However, 1,4-dioxane itself is not used as a cosmetic ingredient." FDA, 1,4-Dioxane, http://www.fda.gov/Cosmetics/ProductandIngredientSafety/ Potential Contaminants/ucm101566.htm.

[118] The Campaign for Safe Cosmetics, What is the Compact for Safe Cosmetics, http://www.safecosmetics.org/ article.php?id=341.

[119] The Campaign for Safe Cosmetics, FAQ: The Compact for Safe Cosmetics, http://www.safecosmetics.org/article.php?id=284#compact.http://www.nottoopretty.org/ article.php?id=284

[120] The Campaign for Safe Cosmetics, No More Toxic Tub: Getting Contaminants Out of Children's Bath & Personal Care Products, http://www.safecosmetics.org/downloads/ NoMore ToxicTub_Mar09Report.pdf.

[121] Environmental Working Group, Skin Deep Database, About the Environmental Working Group, http://www.ewg.org/about.

[122] Keratins are hair proteins. California Department of Public Health, Occupational Health Branch, California Safe Cosmetics Program, Q&A: Brazilian Blowout & Other Hair Smoothing Salon Treatments, http://www.cdph.ca.gov/ programs/cosmetics/Documents/ BrazilianBlowoutQA.pdf.

[123] See the **Appendix** for a more detailed discussion of keratin hair treatment products containing formaldehyde.

[124] E.g., *The Food and Drug Administration's Regulation of Cosmetics: Hearing Before the Subcomm. on Oversight and Investigations of the House Comm. on Interstate and Foreign Commerce* (Feb. 3, 1978) at 4 (statement of Gregory J. Ahart, Director, Human Resources Division, GAO) [hereinafter Ahart Statement]; *The Review of the Adequacy of Existing Laws Designed to Protect the Public from Exposure to Cancer Causing and Other Toxic*

Chemicals in Hair Dyes and Cosmetic Products: *Hearings Before the Subcomm. on Oversight and Investigations of the House Comm. on Interstate and Foreign Commerce*, Serial No. 95-91, 95th Cong. (Jan. 23 and 26, Feb. 2-3, 1978) at 370 (statement of Hon. Donald Kennedy, FDA Commissioner); *Safety of Hair Dyes and Cosmetic Products*: *Hearing Before the Subcomm. on Oversight and Investigations of the H. Comm. on Interstate and Foreign Commerce*, Serial No. 96-105 (July 19, 1979) at 6 (statement of Sherwin Gardner, Acting Commissioner, FDA) [hereinafter Gardner Statement].

[125] 21 C.F.R. § 70.3(f); *see also* FFDCA § 201(t)("(1) The term "color additive" means a material which—(A) is a dye, pigment, or other substance made by a process of synthesis or similar artifice, or extracted, isolated, or otherwise derived, with or without intermediate or final change of identity, from a vegetable, animal, mineral, or other source, and (B) when added or applied to a food, drug, or cosmetic, or to the human body or any part thereof, is capable (alone or through reaction with other substance) of imparting color thereto; except that such term does not include any material which the Secretary, by regulation, determines is used (or intended to be used) solely for a purpose or purposes other than coloring. (2) The term "color" includes black, white, and intermediate grays.").

[126] 21 U.S.C. § 379e(a); FFDCA § 721(a).

[127] 21 U.S.C. § 379e(a); FFDCA § 721(a); 21 U.S.C. § 361(e); FFDCA § 601(e).

[128] "In the certification procedure, a representative sample of a new batch of color additive, accompanied by a 'request for certification' that provides information about the batch, must be submitted to FDA's Office of Cosmetics and Colors. FDA personnel perform chemical and other analyses of the representative sample and, providing the sample satisfies all certification requirements, issue a certification lot number for the batch." 76 Fed. Reg. 10371, 10372 (Feb. 24, 2011); 21 C.F.R. Part 80.

[129] 21 U.S.C. § 379e(a); FFDCA § 721(a); 21 C.F.R. § 71.25.

[130] 21 U.S.C. § 379e(c); FFDCA § 721(c); 21 C.F.R. Part 73, Subpart C, Listing of Color Additives Exempt from Certification; 21 C.F.R. Part 74, Subpart C, Listing of Color Additives Subject to Certification; *see also* FDA, Color Additives Permitted for Use in Cosmetics: Table, http://www.fda.gov/Cosmetics/ GuidanceComplianceRegulatory Information/VoluntaryCosmeticsRegistrationProgramVCRP/OnlineRegistration/ ucm109084.htm.

[131] Ibid.

[132] The FDA's "safe-for-use" principle "require[s] the presentation of all needed scientific data in support of a proposed listing to assure that each listed color additive will be safe for its intended use in or on ... cosmetics." 21 C.F.R. § 70.42. In this context, "safe" means "that there is convincing evidence that establishes with reasonable certainty that no harm will result from the intended use of the color additive." 21 C.F.R. § 70.3(i).

[133] 21 U.S.C. § 379e(a); FFDCA § 721(a).

[134] 21 C.F.R. § 82.304.

[135] 21 C.F.R. § 70.3(v); FDA, Color Additives and Cosmetics, http://www.fda.gov/ForIndustry/ ColorAdditives/ ColorAdditivesinSpecificProducts/InCosmetics/ucm110032.htm

[136] E.g., 21 C.F.R. § 73.2995—Luminescent zinc sulfide; FDA, Color Additives and Cosmetics, *supra* note 121.

[137] 21 C.F.R. § 70.5. The FDA notes that it has not approved any color additive for skin injections such as tattoos or permanent makeup. FDA, Color Additives and Cosmetics, *supra* note 121. Additionally, color additives may be required to be labeled as "Do not use for coloring drugs for injection." 21 C.F.R. § 70.25.

[138] FDA, Color Additives and Cosmetics, *supra* note 135. The FDA also states that coal tar colors are "materials consisting of one or more substances that either are made from coal-tar or can be derived from intermediates of the same identity as coal-tar intermediates," and "may also include diluents or substrata."

[139] FDA, Hair Dye Products, http://www.fda.gov/Cosmetics/ProductandIngredientSafety/Product Information/ ucm143066.htm.

[140] 21 U.S.C. § 361; FFDCA § 601; Hutt, *supra* note 32, p. 7.

[141] *See, e.g., The Food and Drug Administration's Regulation of Cosmetics: Hearing Before the Subcomm. on Oversight and Investigations of the House Comm. on Interstate and Foreign Commerce* (Feb. 3, 1978) at 4 (statement of Gregory J. Ahart, Director, Human Resources Division, GAO) [hereinafter Ahart Statement]; *The Review of the Adequacy of Existing Laws Designed to Protect the Public from Exposure to Cancer Causing and Other Toxic Chemicals in Hair Dyes and Cosmetic Products*: *Hearings Before the Subcomm. on Oversight and Investigations of the House Comm. on Interstate and Foreign Commerce,* Serial No. 95-91, 95th Cong. (Jan. 23 and 26, Feb. 2-3, 1978) at 370 (statement of Hon. Donald Kennedy, FDA Commissioner); *Safety of Hair Dyes and Cosmetic Products*: *Hearing Before the Subcomm. on Oversight and Investigations of the H. Comm. on Interstate and Foreign Commerce*, Serial No. 96-105 (July 19, 1979) at 6 (statement of Sherwin Gardner, Acting Commissioner, FDA) [hereinafter Gardner Statement].

[142] *See, e.g., The Review of the Adequacy of Existing Laws Designed to Protect the Public from Exposure to Cancer Causing and Other Toxic Chemicals in Hair Dyes and Cosmetic Products*: *Hearings Before the Subcomm. on Oversight and Investigations of the House Comm. on Interstate and Foreign Commerce,* 95th Cong., Serial No. 95-91 (Jan. 23 and 26, Feb. 2-3, 1978) at 370 (statement of Hon. Donald Kennedy, FDA Commissioner)("But our ability to protected the public, particularly from the risk associated with long-term use of hair dyes, will continue to be severely limited until Congress repeals the exemptions for coal tar hair dye products. We have long stated that the coal tar exemptions of section 601(a) and (e) and 602(e) should be repealed."); Ahart Statement, *supra* note 124, at 4; Gardner Statement, *supra* note 129, at 6 ("The law does contain an exemption for coal tar hair dyes from the principal adulteration provisions of the act. ... We have long urged that this outmoded exemption be eliminated, and the Department will shortly submit legislation that will accomplish this purpose.").

[143] Ahart Statement, *supra* note 141, p. 4.

[144] FDA, Hair Dye Products, *supra* note 139.

[145] 21 C.F.R. § 740.18; see 47 Fed. Reg. 7829 (Feb. 23, 1982), which stayed this regulation until further notice, effective Sept. 18, 1980; FDA, Hair Dye Products, *supra* note 124.

[146] 21 U.S.C. § 361; FFDCA § 601; 21 C.F.R. § 70.3(u).

[147] 21 U.S.C. § 361(a); FFDCA § 601(a). In addition to the warning label, coal tar hair dyes must have "adequate directions for preliminary patch testing" to meet the exemption from the FFDCA § 601(a) adulteration provisions. 21 C.F.R. § 70.3(u).

[148] National Nanotechnology Initiative (NNI), What is Nanotechnology?, http://www.nano.gov/ nanotech-101/what/definition. According to the NNI, "There are 25,400,000 nanometers in one inch." NNI, Size of the Nanoscale, http://www.nano.gov/nanotech-101/what/nano-size.

[149] Nanotechnology has been used in clear sunscreens and "deep-penetrating therapeutic cosmetics" since 1999 to the early 2000s. NNI, Nanotechnology Timeline, http://www.nano.gov/nanotech-101/timeline.

[150] Johnson, *supra* note 48, p. 88.

[151] Lori McGroder, Shook, Hardy & Bacon, *Nanotechnology – Keeping Cosmetics Out of the Courtroom,* Apr. 7, 2011, http://www.cosmeticsbusiness.com/technical/article_page/ Nanotechnology__keeping_cosmetics_out_of_the_courtroom/60306.

[152] ObservatoryNANO, General Section Reports, Nanotechnology in Cosmetics, at 6.2, http://www.observatorynano.eu/project/filesystem/files/Cosmetics%20report-April%2009. pdf.

[153] The Project on Emerging Nanotechnologies, About Us, Mission, http://www. nanotechproject.org/about/mission/. *See* http://www.nanotechproject.org/ inventories/ consumer/ for the searchable database of consumer products reportedly containing nanomaterials.

[154] The Project on Emerging Nanotechnologies, Health and Fitness, Cosmetics, http://www.nanotechproject.org/inventories/consumer/browse/categories/health_fitness/cos metics/.

[155] Abramowitz, *supra* note 61, p. 203-04.

[156] Albert C. Lin, *Size Matters: Regulating Nanotechnology*, 31 Harv. Envtl. L. Rev. 349, 359, 2007.

[157] Abramowitz, *supra* note 61, p. 207.

[158] Ibid. at 203, 207.

[159] E.g., Friends of the Earth, Nanomaterials, Sunscreens and Cosmetics: Small Ingredients, Big Risks (May 2006). The industry responded to this chapter in a white paper. Johann Wiechers, Nanotechnology and Skin Delivery: Infinitely Small or Infinite Possibilities? 124 Cosmetics & Toiletries Magazine, Jan. 2009, http://www.CosmeticsandToiletries.com; *see also* International Center for Technology Assessment, Citizens Petition to the United States Food and Drug Administration Requesting FDA Amend its Regulations for Products Composed of Engineered Nanoparticles Generally and Sunscreen Drug Products Composed of Engineered Nanoparticles (May 16, 2006), http://www.icta.org/global/actions.cfm?page= 15&type+364&topic_8 (petition regarding regulation of products with unlabeled nanomaterials and their health and environmental risks); Letter to Andrew C. von Eschenbach, FDA Commissioner, from Consumers Union, Oct. 8, 2008, http://www.consumersuiion.org/pub/core_product_safety/ 006254.html (requesting a safety assessment of the use of engineered nanoparticles, particularly in cosmetics, sunscreens, and sunblocks).

[160] Press Release, FDA, FDA Forms Internal Nanotechnology Task Force, Aug. 9, 2006, http://www.fda.gov/ NewsEvents/Newsroom/PressAnnouncements/2006/ucm108707.htm.

[161] FDA, Considering Whether an FDA-Regulated Product Involves the Application of Nanotechnology, Draft Guidance for Industry, at n.4 (June 14, 2011), 76 *Federal Register* 34715, http://www.fda.gov/regulatoryinformation/ guidances/ucm257698.htm [hereinafter Draft Guidance].

[162] FDA, FDA Regulation of Nanotechnology Products, http://www.fda.gov/ScienceResearch/ SpecialTopics/ Nanotechnology/NanotechnologyTaskForce/ucm115441.htm.

[163] FDA, Nanotechnology Task Force, Nanotechnology: A Report of the U.S. Food and Drug Administration, July 25, 2007, http://www.fda.gov/downloads/ScienceResearch/Special Topics/Nanotechnology/ucm110856.pdf.

[164] Ibid. at 14, 16.

[165] Ibid. at 15-16, 30, 32.

[166] FDA, Draft Guidance, *supra* note 162, also available at http://www.gpo.gov/fdsys/pkg/FR-2011-06-14/pdf/2011- 14643.pdf. The draft guidance was issued the same day as a White House memorandum to executive branch departments and agencies on policy principles

regarding U.S. regulation and oversight of nanotechnology and nanomaterials. Memorandum from John P. Holdren, Director, Office of Science and Technology Policy, et al., to the Heads of Executive Departments and Agencies, Policy Principles for the U.S. Decision-Making Concerning Regulation and Oversight of Applications of Nanotechnology and Nanomaterials, June 9, 2011.

[167] FDA, Draft Guidance, *supra* note 162.

[168] Ibid.

[169] Ibid.

[170] FDA, Draft Guidancc for Industry: Safety of Nanomaterials in Cosmetic Products; Availability, 77 *Federal Register* 24722, April 25, 2012, http://www.fda.gov/Cosmetics/ GuidanceComplianceRegulatoryInformation/ GuidanceDocuments/ucm300886.htm.

[171] Ibid.

[172] Ibid.

[173] 21 C.F.R. Part 710—Voluntary Registration of Cosmetic Product Establishments. In May 2008, an estimated one-third of cosmetic establishments were registered. *Discussion Draft of the 'Food and Drug Administration Globalization Act' Legislation: Device and Cosmetic Safety, Hearing Before the Subcomm. on Health, H. Comm. on Energy and Commerce*, 110th Cong., May 14, 2008 (statement of Stephen Sundlof, FDA's Center for Food Safety and Applied Nutrition).

[174] GAO, Cosmetics Regulation: Information on Voluntary Actions Agreed to by FDA and the Industry, Report to the Chairman, Subcommittee on Regulation, Business Opportunities, and Energy, House Committee on Small Business, GAO/HRD-90-58, at 3 (Mar. 1990). The GAO report also commented on a "major disagreement" between FDA and the cosmetic industry's trade group as to the number of companies that were not registered with FDA and stated that FDA's inability to require registration inhibited the agency's ability to "accurately assess how many companies may be avoiding registration." Ibid. pp. 3-4.

[175] FDA, Voluntary Cosmetic Registration Program (VCRP), http://www.fda.gov/Cosmetics/ GuidanceComplianceRegulatoryInformation/VoluntaryCosmeticsRegistrationProgramVCR P/default.htm.

[176] 21 C.F.R. §§ 710.1, 710.2.

[177] 21 C.F.R. § 710.1.

[178] 21 C.F.R. § 720.4.

[179] 21 C.F.R. § 710.1.

[180] 21 C.F.R. § 710.9.

[181] 21 C.F.R. § 710.9.

[182] "FDA disagrees with the suggested change to its registration program. Cosmetic products marketed in the United States are regulated by FDA in accordance with the requirements of the [FFDCA] and, if offered for sale as consumer commodities, the [FPLA]. The FPLA defines a consumer commodity as a product distributed through retail sales for consumption by individuals. Professional products used in salons, and free samples are not available through retail sale to consumers, so they are not considered to be in 'commercial distribution.' Because the VCRP program only applies to cosmetic products in commercial distribution as defined in the FPLA, FDA is unable to file professional cosmetic products." Letter from Thomas Cluderay, Staff Attorney and Stabile Fellow, Environmental Working Group, EWG Comments on FDA's Voluntary Cosmetic Registration Program, Docket No. FDA-2010-N-0623 (Feb. 11, 2011); *see also* Alaina Busch, *Mandatory Cosmetics Adverse Events Reporting Urged*, FDA Week, Apr. 15, 2011.

[183] FDA, Agency Information Collection Activities; Submission for Office of Management and Budget Review; Comment Request; Voluntary Cosmetic Registration Program, 76 Fed. Reg. 10607, 10608 (Feb. 25, 2011).

[184] Ibid.; see also Alaina Busch, *FDA Rejects Cosmetics Certification Program Recommendation*, FDA Week, Mar. 3, 2011.

[185] 76 Fed. Reg. at 10608.

[186] Ibid.

[187] FDA, MedWatch: The FDA Safety Information and Adverse Event Reporting Program, http://www.fda.gov/Safety/ MedWatch/default.htm.

[188] FDA, Bad Reaction to Cosmetics? Tell FDA, http://www.fda.gov/ForConsumers/ ConsumerUpdates/ ucm241820.htm.

[189] Ibid. Data to be reported to the FDA include "the name and contact information for the person who had the reaction; the age, gender, and ethnicity of the product's user; the name of the product and manufacturer; a description of the reaction—and treatment, if any; the healthcare provider's name and contact information, if medical attention was provided; and when and where the product was purchased."

[190] Ibid.

[191] FDA, Tattoos & Permanent Makeup, http://www.fda.gov/Cosmetics/Productand Ingredient Safety/ProductInformation/ucm108530.htm.

[192] FDA, Temporary Tattoos & Henna/Mehndi, http://www.fda.gov/Cosmetics/Productand IngredientSafety/ProductInformation/ucm108569.htm.

[193] Ibid.

[194] FDA, Tattoos & Permanent Makeup, *supra* note 191.

[195] The reports date from Sept. 29, 2008 to March 1, 2011. CIR, Final Amended Report: Formaldehyde and Methylene Glycol, 11 (Oct. 12, 2011), http://www.cir-safety.org/staff_files/Formal10122011final.pdf; FDA Receives Complaints Associated With the Use of Brazilian Blowout, May 24, 2011, http://www.fda.gov/Cosmetics/ ProductandIngredientSafety/ProductInformation/ucm228898.htm.

[196] Press Release, U.S. Department of Labor, US Labor Department's OSHA Issues Hazard Alert to Hair Salon Owners, Workers, Apr. 11, 2011, http://www.osha.gov/pls/oshaweb/ owadisp.show_document?p_table= NEWS_RELEASES&p_id=19584; Press Release, Illinois Dep't of Financial and Professional Regulation, Hair Smoothing Products Might Cause Health Risk, June 3, 2011, http://www.idfpr.com/NEWSRLS/2011/ 06032011Hair SmoothingAlert.asp.

[197] Cosmeticsinfo.org, Consumer Commitment Code, http://www.cosmeticsinfo.org/fdapartner_ ccc2.php#2. This website is sponsored by PCPC and its members. Cosmeticsinfo.org, About This Website and Its Sponsors, http://www.cosmeticsinfo.org/aboutus.php.

[198] Under 21 C.F.R. § 314.80(a), "serious adverse drug experiences" include "death, a life-threatening adverse drug experience, inpatient hospitalization ... a persistent or significant disability/incapacity, or a congenital anomaly/birth defect," as well as important medical events that "may require medical or surgical intervention to prevent one of the outcomes listed in this definition." An "unexpected adverse drug experience" is "[a]ny adverse drug experience that is not listed in the current labeling for the drug product," or an "adverse drug experience that has not been previously observed."

[199] Cosmeticsinfo.org, Consumer Commitment Code, http://www.cosmeticsinfo.org/fdapartner_ ccc3.php.

[200] FDA, "Organic" Cosmetics, http://www.fda.gov/Cosmetics/ProductandIngredientSafety/ ProductInformation/ ucm203078.htm.

[201] United States Department of Agriculture (USDA), Agricultural Marketing Service, National Organic Program (NOP), Cosmetics, Body Care Products, and Personal Care Products, http://www.ams.usda.gov/AMSv1.0/getfile? dDocName=STELPRDC5068442.

[202] Natasha Singer, *Natural, Organic Beauty*, N.Y. Times, p. G1, Nov. 1, 2007. The lack of an FDA definition for these and similar terms may allow a cosmetic manufacturer to make such claims on "a synthetic-based shampoo with one plant derivative" as well as "a synthetic-free face powder formulated with only minerals."

[203] FDA, Center for Food Safety and Applied Nutrition, Office of Cosmetics and Colors, How Smart Are You About Cosmetics? Question 6a, http://www.accessdata.fda.gov/videos/CFSAN/costf/costf-6.html; Singer, *supra* note 207 (noting that "representatives for the government and the beauty industry, as well as some environmental activists, acknowledge that there is no published scientific proof to support the notion that plant-based cosmetics are safer, healthier or more effective for people").

[204] FDA, "Organic" Cosmetics, http://www.fda.gov/Cosmetics/ProductandIngredient Safety/ProductInformation/ucm203078.htm. The FDA also maintains a database of poisonous plants. FDA, FDA Poisonous Plant Database, http://www.accessdata.fda.gov/scripts/plantox/index.cfm.

[205] Singer, *supra* note 202 (quoting Dr. Linda M. Katz, Director of the FDA's Office of Cosmetics and Colors).

[206] Ibid. (quoting Dr. Linda M. Katz, Director of the FDA's Office of Cosmetics and Colors).

[207] 7 C.F.R. Part 205.

[208] 7 C.F.R. § 205.2; USDA, *supra* note 201.

[209] USDA, *supra* note 203.

[210] Singer, *supra* note 202.

[211] 7 C.F.R. §§ 205.303-05; USDA, *supra* note 203.

[212] USDA, NOP, National Organic Standards Board, Certification, Accreditation, and Compliance Committee, Recommendation: Solving the Problem of Mislabeled Organic Personal Care Products, Aug. 30, 2009, http://www.ams.usda.gov/AMSv1.0/getfile? dDocName=STELPRDC5079488.

[213] Ibid.

[214] Renpure Organics, Frequently Asked Questions, http://www.renpure.com/organic-product-questions.html.

[215] USDA, NOP, *supra* note 212.

[216] Ibid.

[217] USDA, NOP, National Organic Standards Board (NOSB), http://www.ams.usda.gov/AMSv1.0/ ams.fetchTemplateData.do?template=TemplateQ&navID=National Organic Program&leftNav=NationalOrganicProgram&page=NOSBHome&description=NOSB& acct=nosb.

[218] NSF International is a non-governmental, "not-for-profit, standards development and testing/certification organization" that provides education in the field of public health and safety and serves manufacturers operating in 80 countries. *See* NSF, About NSF, http://www.nsf.org/business/about_NSF/; NSF, Q&A on the American National Standard for Personal Care Products Containing Organic Ingredients, http://www.nsf.org/business/newsroom/press_releases/documents/110620_Contains-Organic-Ingredients-QA.pdf.

[219] ANSI is "a private, non-profit organization that administers and coordinates the U.S. voluntary standardization and conformity assessment system. ANSI standards are developed on the principles of due-process, participation, and consensus." Q&A on the American National Standard for Personal Care Products Containing Organic Ingredients,

http://www.nsf.org/business/newsroom/press_releases/documents/110620_Contains-Organic-Ingredients-QA.pdf.

[220] NSF/ANSI 305 Personal Care Standard, NSF's New "Contains Organic Ingredients" Standard for Personal Care Products Adopted as American National Standard, http://www.nsf.org/ business/newsroom/articles/ 0903_n3_nsf305.asp.

[221] NSF, Q&A, *supra* note 218.

[222] NSF, Q&A, *supra* note 218.

[223] NSF/ANSI 305 Personal Care Standard, NSF's New "Contains Organic Ingredients" Standard for Personal Care Products Adopted as American National Standard, http://www.nsf.org/ business/newsroom/articles/ 0903_n3_nsf305.asp.

[224] NSF, Q&A, *supra* note 218.

[225] Ibid.

[226] FDA, Cruelty Free/Not Tested on Animals, http://www.fda.gov/Cosmetics/Cosmetic Labeling LabelClaims/ LabelClaimsandExpirationDating/ucm2005202.htm.

[227] Ibid.

[228] Dana Canedy, *P.&.G. to End Animal Tests for Most Consumer Goods*, N.Y. Times, July 1, 1999, http://www.nytimes.com/1999/07/01/business/p-g-to-end-animal-tests-for-most-consumer-goods.html?scp=6&sq=cosmetics%20companies%20quietly%20ending%20 animal%20tests&st=cse.

[229] PETA, PETA's Caring Consumer Program, http://www.peta.org/about/learn-about-peta/caring-consumerprogram.aspx. As another example, the Coalition for Consumer Information on Cosmetics' (CCIC) Leaping Bunny Program allows cosmetics products that meet certain criteria for non-animal tested cosmetic products to bear a "leaping bunny" logo. For this program, the company makes voluntary guarantees regarding the company's and supplier's commitment not to test on animals, and the CCIC may require an independent audit. Coalition for Consumer Information on Cosmetics, The Corporate Standard of Compassion for Animals ("the Standard"), http://www.leaping bunny.org/pdf/Corporate_Standard_of_Compassion_for_Animals.pdf. The independent audit is commissioned either by the company or the CCIC, depending on the company's gross annual sales, and is performed by an accredited auditing firm. Ibid. at 2.

[230] 21 C.F.R. § 701.3.

[231] 15 U.S.C. § 1459(a), 1453.

[232] GAO report, *supra* note 175, at 13.

[233] 21 U.S.C. § 362; FFDCA § 602; 15 U.S.C. § 1459(a). The FPLA limits "consumer commodities" to those products sold at retail.

[234] CIR, Final Amended Report, *supra* note 196, at 6.

[235] First Research, Industry Profile, Cosmetics, Beauty Supply, and Perfume Stores, May 23, 2011.

[236] Ibid.; GAO report, *supra* note 175, pp. 13-14; 15 U.S.C. §§ 1452(a), 1456(a).

[237] OSHA, Hazard Alert: Hair Smoothing Products That Could Release Formaldehyde, http://www.osha.gov/SLTC/ formaldehyde/hazard_alert.html.

[238] CIR, Final Amended Report, *supra* note 196, p. 7.

[239] Keratins are hair proteins. California Department of Public Health, Occupational Health Branch, California Safe Cosmetics Program, Q&A: Brazilian Blowout & Other Hair Smoothing Salon Treatments, http://www.cdph.ca.gov/ programs/cosmetics/Documents/ BrazilianBlowoutQA.pdf

[240] OSHA, OSHA Fact Sheet: Formaldehyde, http://www.osha.gov/OshDoc/data_General_Facts/ formaldehydefactsheet.pdf. NIOSH is an entity within CDC.

[241] Hair salons and stylists are regulated at the state level. States may require salons and barber shops to register with the state and stylists to apply for and possess a license.

[242] Letter to Dr. Margaret A. Hamburg, M.D., Commissioner of Food and Drugs, FDA, from Reps. Schakowsky, Markey, Baldwin, Blumenauer, Conyers, Lowey, Moran, Lee, Chu, and Deutch, May 6, 2011, http://schakowsky.house.gov/images/stories/ Letter_to_FDA_on_ Dangerous_Chemicals_in_Brazilian_Blowout_Hair_Treatments.pdf. Hair salons and stylists are regulated at the state level. States may require salons and barber shops to register with the state and stylists to apply for and possess a license.

[243] Letter from Michael W. Roosevelt, Acting Director, Office of Compliance, Center for Food Safety and Applied Nutrition, FDA, to Mike Brady, CEO, GIB, LLC, dba Brazilian Blowout, Aug. 22, 2011, http://www.fda.gov/ICECI/ EnforcementActions/ Warning Letters/ucm270809.htm.

[244] FDA, Cosmetics: FDA, OSHA Act on Brazilian Blowout, Oct. 21, 2011, http://www.fda.gov/ Cosmetics/ ProductandIngredientSafety/ProductInformation/ucm228 898.htm.

[245] Brazilian Blowout, Brazilian Blowout Now Working Directly with the FDA to Help Clear Up the Controversy, Sept. 19, 2011, http://www.brazilianblowout.com/fda.

[246] Letter from Frank Meilinger, Director, Office of Communication, Assistant Secretary for Occupational Safety and Health, U.S. Department of Labor, to Michael Brady, CEO, Oill LLC dba Brazilian Blowout, Sept. 22, 2011, http://osha.gov/SLTC/formaldehyde/ brazilian _blowout_letter.html.

[247] EPA, An Introduction to Indoor Air Quality (IAQ), Formaldehyde, http://www.epa.gov/iaq/ formaldehyde.html#Health Effects. EPA is conducting a risk assessment for formaldehyde in order to update information in its Integrated Risk Information System (IRIS), a database of chemical toxicity information. A 1991 IRIS assessment classified formaldehyde as a probable human carcinogen. EPA currently is revising a draft updated risk assessment based on comments from reviewers at the National Academy of Sciences.

[248] Agency for Toxic Substances & Disease Registry (ATSDR), Toxic Substances Portal - Formaldehyde, ToxFAQs TM for Formaldehyde, http://www.atsdr.cdc.gov/toxfaqs/ TF. asp?id=219&tid=39. ATSDR is an entity within CDC.

[249] HHS, Public Health Service, National Toxicology Program, Report on Carcinogens (12th ed.), http://ntp.nih.gov/go/ roc12.

[250] Agents Classified by IARC Monographs, Volumes 1-102, http://monographs.iarc.fr/ENG/ Classification/ClassificationsAlphaOrder.pdf.

[251] OSHA, Safety and Health Topics, Formaldehyde, http://www.osha.gov/SLTC/formaldehyde/ index.html/.

[252] Ibid.

[253] Ibid.

[254] 29 C.F.R. §1910.1048(a).

[255] 29 C.F.R. § 1910.1048(c).

[256] Ibid.

[257] Ibid.

[258] 29 C.F.R. § 1910.1048(d).

[259] 29 C.F.R. § 1910.1048(m).

[260] 29 C.F.R. § 1910.1048(m)(3).

[261] 29 C.F.R. § 1910.1048(m)(4).

[262] Brazilian Blowout, Frequently Asked Questions, http://www.brazilianblowout.com/faq/; Brazilian Blowout, Material Safety Data Sheet, http://www.brazilianblowout.com/_ literature_72696/Material_Safety_Data_Sheet.pdf.

[263] Katy Muldoon, *Brazilian Blowout Drops Lawsuit Against Oregon OSHA and OSHU*, OregonLive.com (Mar. 2, 2011), http://www.oregonlive.com/health/ index.ssf/2011/03/ brazilian_blowout_drops_lawsui.html; Oregon OSHA, A Division of the Oregon Department of Consumer and Business Services, and CROET at Oregon Health & Sciences University, "Keratin-Based" Hair Smoothing Products and the Presence of Formaldehyde (2010), http://www.orosha.org/pdf/Final_ Hair_Smoothing_Report.pdf.

[264] Oregon OSHA, *supra* note 264, at 1.

[265] Ibid. at 2.

[266] Muldoon, *supra* note 264.

[267] Oregon OSHA, *supra* note 264, p. 2.

[268] OSHA, Hazard Alert: Hair Smoothing Products That Could Release Formaldehyde, http://www.osha.gov/SLTC/ formaldehyde/hazard_alert.html.

[269] CIR, Final Amended Report, *supra* note 196, p. 11. Epiglottitis is inflammation of tissues in the back of the throat.

[270] Ibid. Pneumonitis is inflammation of the lungs.

[271] CIR, Final Amended Report, *supra* note 196, at 11.

[272] Letter from Srinivas Durgam, Industrial Hygienist, and Elena Page, Medical Officer, Hazards Evaluations and Technical Assistance Branch, Division of Surveillance, Hazard Evaluations, and Field Studies, National Institute for Occupational Safety and Health, to Salon Owners (May 16, 2011), http://www.cdc.gov/niosh/hhe/pdfs/HETA_11- 0014_ Interim_Letter_for_web.pdf

[273] Ibid. at 6.

[274] OSHA, Hazard Alert, *supra* note 269.

[275] FDA, FDA Receives Complaints Associated With the Use of Brazilian Blowout, (May 24, 2010), http://www.fda.gov/Cosmetics/ProductandIngredientSafety/ProductInformation/ucm 228898.htm.

[276] Ibid.

[277] Letter from Frank Meilinger, Director, Office of Communication, Assistant Secretary for Occupational Safety and Health, U.S. Department of Labor, to Michael Brady, CEO, Oill LLC dba Brazilian Blowout (Sept. 22, 2011), http://osha.gov/SLTC/formaldehyde/ brazilian_blowout_letter.html.

[278] OSHA, Hazard Alert: Hair Smoothing Products That Could Release Formaldehyde, http://www.osha.gov/SLTC/ formaldehyde/hazard_alert.html.

[279] Cal/OSHA, Hair Smoothing Products that May Contain or Release Formaldehyde, http://www.dir.ca.gov/dosh/ HairSmoothingPageVersion1Nov182010.pdf.

[280] Letter from Frank Meilinger, Director, Office of Communication, Assistant Secretary for Occupational Safety and Health, U.S. Department of Labor, to Michael Brady, CEO, Oill LLC dba Brazilian Blowout, Sept. 22, 2011, http://osha.gov/SLTC/formaldehyde/ brazilian_blowout_letter.html.

[281] Ibid.

[282] Press Release, Health Canada, Several Professional Hair Smoothing Solutions Contain Excess Levels of Formaldehyde (Apr. 12, 2011), http://www.hc-sc.gc.ca/ahc-asc/media/advisories-avis/_2011/2011_56-eng.php.

[283] CIR, Final Amended Report, *supra* note 196, p. 7.

[284] Ibid. p. 17. "[I]n no case should the formalin concentration exceed 0.2% (w/w), which would be 0.074% (w/w) calculated as formaldehyde or 0.118% (w/w) calculated as methylene glycol."

[285] CIR, Final Amended Report, *supra* note 196, p. 7.

[286] CIR, Final Amended Report, *supra* note 196. Neither formaldehyde nor methylene glycol is available commercially, but are "produced as an aqueous solution called formalin."
[287] Ibid. p. 8.
[288] Ibid. p. 17.
[289] Ibid. pp. 17-18.
[290] Ibid. p. 7.
[291] Ibid. p. 3, 18.
[292] CIR, Final Amended Report, supra note 196, pp. 19, 22. This amount "would be 0.074%(w/w) calculated as formaldehyde or 0.118%(w/w) calculated as methylene glycol."
[293] Ibid. at 22.
[294] The Chief Scientist of the Personal Care Products Council has issued a statement recommending that FDA "take prompt and appropriate action to make sure these products have been fully tested and substantiated for safety under their conditions of use." Additionally, the PCPC "strongly advise[d] consumers and beauticians not to use professional hair straightening products in the home," and to ensure that salons that offer such treatments have proper ventilation. Press Release, Statement by John Bailey, Chief Scientist, Personal Care Products Council, on Cosmetic Ingredient Review (CIR) Expert Panel Preliminary Findings on Safety of Two Ingredients Used in Professional Hair Smoothing Products (Mar. 9, 2011), http://www.personalcarecouncil.org/ newsroom/ 20110309.
[295] FDA, Cosmetics: FDA, OSHA Act on Brazilian Blowout (Oct. 21, 2011), http://www.fda.gov/Cosmetics/ProductandIngredientSafety/ProductInformation/ucm228898 .htm.
[296] Letter from Phil Broadbent for Kristina Harper, Supervisory Congressional Affairs Specialist, to The Honorable Earl Blumenauer, House of Representatives, Nov. 26, 2010.
[297] CIR, Final Amended Report, *supra* note 196, p. 6.
[298] Letter from Phil Broadbent, *supra* note 296 (citing FDA, Nail Care Products, http://www.fda.gov/Cosmetics/ProductandIngredientSafety/ProductInformation/ucm1270 68.htm#forma).
[299] 21 C.F.R. § 700.11-700.27.
[300] 21 U.S.C. § 361(a); FFDCA § 601(a).
[301] FFDCA §§ 301, 303.
[302] 21 C.F.R. § 740.1(a).
[303] 21 C.F.R. § 1.21(a)(1).
[304] Letter from Phil Broadbent, *supra* note 296.
[305] Letter from Michael W. Roosevelt, Acting Director, Office of Compliance, Center for Food Safety and Applied Nutrition, FDA, to Mike Brady, CEO, GIB, LLC, dba Brazilian Blowout, Aug. 22, 2011, http://www.fda.gov/ICECI/ EnforcementActions/Warning Letters/ucm270809.htm.
[306] Ibid. (referencing 21 U.S.C. § 361(a)).
[307] Ibid. (referencing 21 U.S.C. § 321(n)).
[308] Ibid.
[309] 15 U.S.C. § 52.
[310] California's Department of Public Health released a question and answer document in response to inquires from hair stylists and customers that indicated that employers were required by California's OSHA standards "to protect their employees from exposure to hazardous airborne chemicals in California workplaces." California Department of Public Health, Occupational Health Branch, California Safe Cosmetics Program, Q&A: Brazilian

Blowout & Other Hair Smoothing Salon Treatments, Mar. 3, 2011, http://www.cdph.ca.gov/ programs/cosmetics/Documents/ BrazilianBlowoutQA.pdf.

[311] Press Release, State of California Department of Justice, Office of the Attorney General, Attorney General Kamala D. Harris Announces Settlement Requiring Honest Advertising over Brazilian Blowout Products, Jan. 30, 2012, http://ag.ca.gov/newsalerts/print_release.php?id=2617.

[312] 21 C.F.R. § 721.

[313] Letter from Phil Broadbent, *supra* note 296.

[314] FDA, Cosmetics, FDA Receives Complaints Associated with the Use of Brazilian Blowout, May 24, 2010, http://www.fda.gov/Cosmetics/ProductandIngredientSafety/ProductInformation/ucm228898.htm.

[315] 21 C.F.R. § 701.3.

[316] 21 U.S.C. § 362; FFDCA § 602; 15 U.S.C. § 1459(a).

[317] Letter from Phil Broadbent, *supra* note 296.

In: Cosmetics and FDA Regulation
Editors: A. Garcia and R. DiBartolo
ISBN: 978-1-62257-892-4

Chapter 2

BACKGROUND MEMORANDUM FOR THE HEARING ON "EXAMINING THE CURRENT STATE OF COSMETICS"*

To: Energy and Commerce Committee Members
From: Majority Staff
Re: Examining the Current State of Cosmetics

On Tuesday, March 27, 2012, at 10:15 a.m. in 2322 Rayburn House Office Building, the Subcommittee on Health will hold a hearing entitled "Examining the Current State of Cosmetics." The following provides background on the hearing.

II. BACKGROUND

The cosmetics industry has been regulated by FDA since the enactment of the Federal Food, Drug and Cosmetic Act of 1938 (FFDCA). Currently, FDA's CFSAN is responsible for regulating cosmetics. Similar to drugs, devices and food, the FFDCA prohibits the introduction of adulterated or misbranded cosmetics into interstate commerce and provides for seizure, criminal penalties and other enforcement authorities for violations of the FFDCA. In addition, under the authority of the Fair Packaging and Labeling

* This is an edited, reformatted and augmented version of hearing given on March 27, 2012 of the House Committee on Energy and Commerce.

Act (FPLA), FDA requires an ingredient declaration for cosmetics to enable consumers to make informed purchasing decisions. Cosmetics that fail to comply with the FPLA are considered misbranded under the FFDCA.

Under current law, FDA cannot require cosmetic facilities to register, but FDA does allow these facilities to do so voluntarily through the Voluntary Cosmetic Registration Program (VCRP). Once a company registers, FDA assigns a registration number to each manufacturing establishment. The VCRP also allows for the filing of a cosmetic product ingredient statement for each product the firm has entered into commercial distribution in the United States. Under the FFDCA, FDA has authority to inspect cosmetic manufacturing facilities.

One organization important to the safety of cosmetics is the Cosmetic Ingredient Review (CIR). According to FDA, the CIR expert panel “is an independent, industry-funded panel of medical and scientific experts that meets quarterly to assess the safety of cosmetic ingredients based on data in the published literature as well as some that is voluntarily provided by the cosmetic industry. The industry data may or may not be complete. FDA takes the results of CIR reviews into consideration when evaluating safety, but the results of FDA safety assessments may differ from those of CIR.” FDA representatives do attend CIR meetings in a non-voting capacity.

In recent years, some States have considered legislation that would affect the ingredients that can be used in cosmetic products. Some groups have called for national standards for ingredients of cosmetic products that are reviewed by the FDA. Given the flow of cosmetic products between States, a uniform standard for cosmetic ingredients would serve to further public health by ensuring these decisions are made using sound science and ensure that the interstate flow of cosmetic products is not disrupted by differing State standards.

The FDA’s resources devoted to cosmetics have increased significantly in recent years. In Fiscal Year 2007, the cosmetic activities at FDA received $3.5 million in funding. In Fiscal Year 2012, the funding increased to $11.2 million.

In: Cosmetics and FDA Regulation
Editors: A. Garcia and R. DiBartolo
ISBN: 978-1-62257-892-4

Chapter 3

STATEMENT OF JOSEPH R. PITTS, CHAIRMAN, THE SUBCOMMITTEE ON HEALTH. HEARING ON "EXAMINING THE CURRENT STATE OF COSMETICS"*

Cosmetics are regulated by FDA under the Federal Food, Drug, and Cosmetic Act (FFDCA) of 1938.

The FFDCA forbids the introduction of adulterated or misbranded cosmetics into interstate commerce and provides for seizure, criminal penalties, and other enforcement authorities for violations of the Act.

The Fair Packaging and Labeling Act (FPLA) also requires cosmetics to carry an ingredient declaration to help consumers make informed purchasing decisions.

Unlike other products regulated by FDA, however, such as drugs, medical devices, and biologics, most cosmetic products and ingredients are not subject to FDA premarket approval. Instead, cosmetic manufacturers are largely responsible for substantiating the safety of their products and ingredients before they go to market.

Currently, cosmetic facilities can register with FDA on a voluntary basis, but FDA cannot compel them to do so. While FDA has the authority under FFDCA to enter and inspect cosmetic manufacturing facilities, the industry does not pay user fees for this purpose.

---

* This is an edited, reformatted and augmented version of testimony given on March 27, 2012 before the House Committee on Energy and Commerce, Subcommittee on Health.

According to a June 2010 study by PriceWaterhouseCoopers, the personal care or cosmetics industry is responsible for 2.8 million jobs in the United States, and small businesses create the vast majority of these positions.

For the past several years, the industry and members of both parties have been reviewing FDA's regulatory authority over these products. One issue under review is the need for a national uniform standard for cosmetic products and preemption of state legislation.

I want to welcome each of our witnesses today, and I hope you can share your perspectives on several matters, including: what deficiencies, if any, you currently see in FDA's regulatory authority over cosmetics; what new authorities, if any, do you believe FDA needs in this area; and if new authorities are needed, what will be the impact on small businesses across the country?

In: Cosmetics and FDA Regulation
Editors: A. Garcia and R. DiBartolo
ISBN: 978-1-62257-892-4

Chapter 4

STATEMENT OF MICHAEL M. LANDA, DIRECTOR, CENTER FOR FOOD SAFETY AND APPLIED NUTRITION, FOOD AND DRUG ADMINISTRATION. HEARING ON "EXAMINING THE CURRENT STATE OF COSMETICS"*

INTRODUCTION

Good afternoon, Mr. Chairman and Members of the Subcommittee. I am Michael Landa, Director of the Center for Food Safety and Applied Nutrition at the Food and Drug Administration (FDA or the Agency), which is part of the Department of Health and Human Services. I am pleased to be here today to discuss FDA's oversight of cosmetics. Every day across the country, Americans—men, women, and children—use a wide variety of cosmetic products, including skin moisturizers, shampoos, perfumes, lipsticks, nail polishes, eye and face make-up, hair colors, and deodorants. These consumers expect their cosmetics—and the wide variety of individual ingredients in these products—to be safe. FDA plays a critical role in ensuring that the nation's cosmetics are among the safest in the world.

* This is an edited, reformatted and augmented version of testimony given on March 27, 2012 before the House Committee on Energy and Commerce, Subcommittee on Health.

In my testimony today, I will describe FDA's current authorities and activities to oversee the safety of cosmetics, the challenges we face due to changes in the industry and the increasingly global marketplace, and the new authorities the Administration is seeking to strengthen FDA's regulatory oversight of cosmetics.

CURRENT AUTHORITIES AND ACTIVITIES RELATED TO COSMETIC SAFETY

The Federal Food, Drug, and Cosmetic Act (FD&C Act) defines a cosmetic as an "article intended to be rubbed, poured, sprinkled, or sprayed on, introduced into, or otherwise applied to the human body or any part thereof for cleansing, beautifying, promoting effectiveness, or altering the appearance." The definition also includes articles intended for use as a component of any such articles. Cosmetics firms are responsible for substantiating the safety of their products and ingredients before marketing. However, they are not required to submit safety substantiation data to the Agency, nor to make it available to the Agency. Under the FD&C Act, cosmetic products and ingredients (with the exception of color additives) are not subject to FDA premarket approval or premarket notification.

In general, except for color additives and those ingredients which are prohibited or restricted from use in cosmetics by regulation, a manufacturer may use any ingredient in a cosmetic, provided that the ingredient does not adulterate the finished cosmetic and the finished cosmetic is properly labeled. FDA regulations prohibit or restrict the use of 10 types of ingredients in cosmetic products due to safety concerns. Some examples are chloroform, methylene chloride, and mercury-containing compounds. If manufacturers do not remove dangerous products from the market once a safety concern emerges, the Agency can pursue enforcement actions against violative products or against firms or individuals who violate the law.

Regulations are in place that specify the labeling requirements for cosmetics. These requirements include:

- An identity statement indicating the nature and use of the product (for example, "shampoo" or "lip gloss");
- The name and place of business of the manufacturer, packer, or distributor;

- A net quantity of contents statement in terms of weight, measure, or numerical count (e.g., "net wt. 4 oz.") to inform consumers of the quantity of the cosmetic in the package;
- Material facts about the product and its use (for example, directions for safe use, if a product could be unsafe if used incorrectly);
- Warning and caution statements for products that are required to bear such statements by the FD&C Act and FDA's regulations (for example, coal tar hair dyes); and
- A list of ingredients, in descending order of predominance.

Cosmetic product labels do not need to provide information on how consumers and health care professionals can report adverse events to the manufacturer, packer, or distributor. However, FDA has long encouraged cosmetics manufacturers and distributors to report adverse events voluntarily.

FDA also encourages companies to register their establishments through the Voluntary Cosmetic Registration Program (VCRP) and file cosmetic product ingredient statements with FDA; however, there is no requirement in the FD&C Act for firms to do either. The Agency established the VCRP and the cosmetic product ingredient statement program to gain more information about cosmetics that are being manufactured and marketed to consumers in the United States. The VCRP currently has almost 1,600 domestic and foreign registered cosmetics establishments, and cosmetic product ingredient statements have been filed for over 39,000 products; however, we estimate that only one-third of cosmetics manufacturers voluntarily file cosmetic product ingredient statements for their products with FDA.

FDA participates in the Cosmetic Ingredient Review (CIR) panel, which was established in 1976 by industry, with the support of FDA and the Consumer Federation of America (CFA). The panel consists of academic experts in the fields of dermatology, pharmacology, toxicology, and chemistry, who are voting members of the panel, as well as three non-voting, liaison representatives from FDA, CFA, and industry. The purpose of CIR is to provide expert review of cosmetic ingredients having potential safety issues. Substances for review are chosen based on frequency of use and safety concerns raised by industry, FDA, or other regulatory bodies within the United States or abroad. Data is compiled by the CIR staff and forwarded to panel members for review and discussion at quarterly meetings, which are open to consumers, industry and the press.

CHALLENGES

During the past several years, Americans have seen a dramatic increase in the numbers and types of cosmetic products on the market. Over 8 billion personal care products, which include primarily cosmetics but also some over-the-counter (OTC) drugs and some products regulated by the Consumer Product Safety Commission, are sold annually in the United States. Estimates of annual U.S. sales of these products range from $54 to over $60 billion. Cosmetic products and ingredients are also entering the United States from a growing number of countries, most of which have regulatory systems and standards that are different from those of the United States. From FY 2004 to FY 2010, the number of cosmetics imports has nearly doubled, growing from less than 1 million import entry lines[1] in FY 2004 to more than 1.9 million import entry lines in FY 2010. We expect this upward trend in imported cosmetics and cosmetic ingredients to continue.

To help address this challenge, FDA and its counterparts in the European Union, Canada, and Japan established a forum in 2007 to exchange ideas and better align practices for maintaining global consumer protection in the cosmetics arena without creating unnecessary obstacles to international trade. The forum, known as the International Cooperation on Cosmetics Regulation (ICCR), meets annually to discuss topics of mutual interest in which cooperation may be possible. The meetings include opportunities for participation by representatives from the cosmetics industry and non-governmental organizations. This year, the United States Government is hosting the annual ICCR meeting July 10-13 in Rockville, Maryland. FDA is working with other ICCR regulatory authorities to hold a stakeholder session with organizations active in the field of cosmetics as well as regulatory officials from additional countries who have expressed an interest in participating in this activity. The session will provide an opportunity for the exchange of viewpoints among a broad range of participants and may identify potential areas for future activities and further alignment. FDA is holding a public meeting on May 15 in advance of the ICCR annual meeting to solicit information, such as agenda topics, from interested parties. Since 2007, ICCR has developed principles for addressing cosmetic Good Manufacturing Practices and working documents to address characterization of nanomaterials, and formed a group to address alternatives to animal testing. ICCR continues to work on a variety of other issues related to cosmetics safety and regulation.

In addition to the challenges posed by an increasingly global marketplace, the cosmetics industry is rapidly undergoing significant changes as the technologies used in manufacturing become increasingly sophisticated and the ingredients more complex. The use of nanotechnology may result in cosmetic products or ingredients with different chemical or physical properties than their counterparts that do not contain nanomaterials. Properties and phenomena emerging at the nanoscale may alter the safety, effectiveness, performance, or quality of products—giving rise to both risks and benefits. For example, FDA is conducting research on the ability of different types of nanoscale particles to penetrate skin and on the potential phototoxicity of nano-sized metal oxides used in topical cosmetics. Nanotechnology is an emerging area of science, where there is a critical need to learn more about the potential safety impact.

FDA continues to be actively involved in the National Nanotechnology Initiative, one of the largest federal interagency research and development initiatives, which coordinates funding for nanotechnology research and development among the 26 participating federal departments and agencies. In addition, FDA has a Nanotechnology Task Force to help assess questions regarding FDA's regulatory authorities as they relate to nanotechnology. Through the work of FDA's task force, last June FDA released a Draft Guidance for Industry entitled "*Considering Whether an FDA-Regulated Product Involves the Application of Nanotechnology*" to help industry and others identify when they should consider potential implications for regulatory status, safety, effectiveness, or public health impact that may arise with the application of nanotechnology in FDA-regulated products. The Agency is developing draft guidance for industry on FDA's current thinking on the safety assessment of nanotechnology in cosmetics.

The category of products that straddles the line between cosmetics and drugs also presents new regulatory challenges. The industry often refers to these products as "cosmeceuticals," a term which has no legal or regulatory definition in the United States. This class of products presents new regulatory challenges in a number of ways, including how such products should be regulated and with what requirements such products should comply. Many products in this category are advertised as containing "active ingredients," which, by virtue of the ingredients themselves or the claims made for the product, may cause the product to be classified under the FD&C Act as a drug. The use of such ingredients is increasing, and we expect this trend to continue, posing additional regulatory challenges. For example, retinol, an ingredient used in cosmetic anti-wrinkle preparations (as well as OTC drug preparations), was not listed in any cosmetic product ingredient statement in FDA's Voluntary Cosmetic Registration database prior to 2005 but, by the end of

2006, it was listed in 68. It is currently listed in 200 cosmetic product ingredient statements. Peptides, a class of cosmetic ingredient also used in skin-care preparations and associated with certain drug-like product claims, were not listed in any cosmetic product ingredient statements filed with FDA prior to 2005. Currently, there are over 95 different peptides listed in a total of over 1,200 cosmetic product ingredient statements.

FY 2013 President's Budget

In response to the challenges noted earlier, and to ensure adequate oversight of cosmetics, the FY 2013 President's Budget request includes new legislative authority for FDA to require domestic and foreign cosmetics manufacturers to register with FDA and pay an annual registration fee. The user fees would support FDA's cosmetics safety and other cosmetics-related responsibilities and are estimated to generate $19 million in new resources. The product, ingredient, and facility information submitted with registration would expand FDA's information about the industry and better enable the Agency to develop necessary guidance and safety standards. It would also enable the Agency to identify and address research gaps, for example, about the safety of novel ingredients. With these additional funding resources, FDA would be able to conduct priority activities that meet public health and industry goals.

Specifically, the Agency would conduct the following activities with the new user fee resources:

- Establish and maintain a mandatory Cosmetic Registration Program;
- Acquire, analyze, and apply scientific data and information from a variety of sources, including voluntary adverse event reporting, to set U.S. cosmetics safety standards;
- Maintain a strong U.S. presence in international standard-setting efforts;
- Provide education, outreach, and training to industry and consumers, and
- Refine inspection and sampling of domestic and imported products and apply risk-based approaches to post-market monitoring of domestic and imported products and other enforcement activities.

Overall, the new authority for registration and user fees would strengthen FDA's ability to protect American consumers from potentially unsafe cosmetic products or ingredients.

Conclusion

FDA is committed to ensuring the safety of cosmetics used by consumers across the United States. The Agency will continue to work closely with all of its partners on a wide variety of issues important to ensuring cosmetics safety. As Congress considers potential steps to address these issues, we look forward to working with you.

Thank you for the opportunity to discuss FDA's activities to ensure the safety of cosmetics. I would be happy to answer any questions you may have.

End Note

[1] An import entry line is a portion of an import entry that is listed as a separate item on an entry document. An importer may identify merchandise in an entry in multiple portions; however, an item in the entry having a different tariff description must be listed separately.

In: Cosmetics and FDA Regulation
Editors: A. Garcia and R. DiBartolo
ISBN: 978-1-62257-892-4

Chapter 5

TESTIMONY OF HALYNA BRESLAWEC, CHIEF SCIENTIST, PERSONAL CARE PRODUCTS COUNCIL. HEARING ON "EXAMINING THE CURRENT STATE OF COSMETICS"*

Chairman Pitts, Ranking Member Pallone and distinguished Members of the Committee, thank you for the opportunity to testify before you on behalf of the Personal Care Products Council. My name is Halyna Breslawec. I am the Chief Scientist and Executive Vice President for Science for the Personal Care Products Council and hold a PhD in Medicinal Chemistry. Prior to joining the Council, I spent 14 years at the U.S. Food and Drug Administration (FDA,) worked in the private sector as a medical device consultant, and served as the deputy director of the Cosmetic Ingredient Review or CIR, an independent body of experts that assesses the safety of ingredients used in cosmetics in the U.S. I am here today to speak about the important role that science plays in the cosmetics industry.

Cosmetics are among the safest category of products regulated by the FDA. The safety of our consumers and their families is always the number one priority for our industry. Careful and thorough scientific research and development are the most important aspects of cosmetic formulation and the foundation for everything that we do. The American cosmetics industry

---

* This is an edited, reformatted and augmented version of testimony given on March 27, 2012 before the House Committee on Energy and Commerce, Subcommittee on Health.

invests more than $3.6 billion each year on scientific research and development. As a result of this research, 2,000 new products are launched each year, and numerous scientific studies are published on enhancing or developing new safety methods.

A regulatory structure should be comprehensive and robust, but should not be so overly burdensome that it stifles or prevents companies from delivering innovative products to the marketplace.

Product safety is a priority for each of our member companies and for our trade association. The companies we represent invest substantial resources each year to ensure the safety and efficacy of their products. Companies work diligently with chemists, toxicologists, microbiologists, dermatologists, environmental scientists and other scientific experts to evaluate and ensure the safety of cosmetic products before they reach the marketplace.

Companies conduct product safety evaluations using the same science-based approaches embedded in FDA, EPA, and other regulatory agencies around the world. Cosmetic safety assessments are thorough and address numerous health questions, including, but not limited to the potential for cancer, reproductive harm, allergy, and how an ingredient is cleared if it reaches the body. The foundation of science-based safety assessment is that any ingredient has a safe range and an unsafe range whether it is water, or a vitamin, or a newly discovered compound. An ingredient's safe range is defined through many, many studies before it can be used in a product. Safety is about choosing ingredients that can be used well within their safe range and avoiding ingredients that cannot be used safely. A complete safety assessment also accounts for who uses the products, how they are used and how often, over a lifetime. Finally, companies' post market surveillance of the consumer experience acts to affirm product safety.

In addition to the work of each individual company, our trade association supports outside, independent programs to review product and ingredient safety. Perhaps the most significant example of this is the Cosmetic Ingredient Review or CIR, which was established in 1976 with support from the FDA and the Consumer Federation of America.

Today, CIR is the only scientific program in the world dedicated to a thorough and continuous review of cosmetic ingredient safety in a public forum. The CIR Expert Panel is an independent, non-profit body of world-renowned physicians and scientists who examine and assess cosmetic ingredient safety data in an open, public manner. Their work is critical to our industry. The FDA and the Consumer Federation of America, along with the Council, serve as non-voting members of CIR and play a valuable role in the deliberations. CIR has reviewed the safety of more than 2,400 cosmetic ingredients and publishes its findings in a transparent manner. These reviews define safe ranges for ingredients used in products, and each ingredient report often involves the panel's scrutiny of hundreds of studies. CIR has also evaluated the safety of certain cosmetic ingredients at the request of FDA.

Consumer, scientific and medical groups nominate the CIR Expert Panel members who must meet strict conflict of interest standards. Just as important, CIR maintains a completely open and transparent process – all CIR meetings are open to the public, as is all of the safety data that they evaluate. Members of the public can also raise issues to be included on the agenda for panel meetings. CIR's findings are published in the peer-reviewed scientific journal, *The International Journal of Toxicology*.

We strongly recommend that FDA incorporate the CIR into its product regulatory process. FDA should formally recognize the findings of the CIR Expert Panel as part of the regulatory regime for cosmetics. Science and safety are the foundation of the cosmetic industry and collectively we must remain steadfast in our commitment to safety. Acceptance and reliance on CIR findings will affirm that commitment.

I'd like to take off my science hat for a moment, and on behalf of the Council, say a few words about the enormous contributions our industry is making to the U.S. economy, specifically to small businesses and what we see is at stake here.

The cosmetic industry plays a unique role in the lives of American women, and not just as women consumers. Our industry is committed to enhancing their lives in a number of ways. We are dedicated to ensuring women have advantages and opportunities for both their professional and personal success. Women comprise 66% of our industry's workforce, compared to 48% of the overall workforce.

Women now hold more than half of all management positions in our industry, compared with 36% of industry in general. Moreover, women of color represent 22% of our total workforce, and 11% of management, compared to 17% of employment and 7% of management industries throughout the entire economy.

Council member companies that are direct sellers like Avon, Mary Kay, Herbalife and Amway, among others, offer strong entrepreneurial opportunities for women across America – opportunities that allow for personal growth and economic freedom.

Chairman Pitts, Ranking Member Pallone, and distinguished members of the committee, thank you again for the opportunity to testify today. The cosmetic industry puts consumer safety first, and we will continue to proactively work to ensure the products we manufacture contribute to the well-being of American consumers. Our work and that of our members is based on sound scientific principles. We look forward to working with you and your staff to modernize FDA's cosmetic regulatory structure so that the agency can act as effectively as it needs to provide peace of mind to the women and men who use our products. This will also give businesses the certainty they need to continue to innovate and provide consumers access to both the legacy brands and the new, exciting and safe products they have come to expect.

Thank you.

In: Cosmetics and FDA Regulation
Editors: A. Garcia and R. DiBartolo
ISBN: 978-1-62257-892-4

Chapter 6

TESTIMONY OF PETER BARTON HUTT, SENIOR COUNSEL, COVINGTON AND BURLING, LLP. HEARING ON "EXAMINING THE CURRENT STATE OF COSMETICS"*

Mr. Chairman, Ranking Member Pallone, and Members of the Committee, I am Peter Barton Hutt. I am Senior Counsel at the Washington, D.C. law firm of Covington and Burling, and a Lecturer on Food and Drug Law at Harvard Law School where I have taught a course on Food and Drug Law for the past 19 years. During 1971 - 1975, I served as Chief Counsel for the Food and Drug Administration.

Thank you for the opportunity to appear before you today on behalf of the Personal Care Products Council, the trade association representing the cosmetic industry in the United States and globally. With me are Dr. Halyna Breslawec, Chief Scientist and Executive Vice President for Science at the Personal Care Products Council, and Ms. Curran Dandurand, CEO and Co-Founder of Jack Black Skincare, a Texas based small business. Ms. Dandurand is here on behalf of the Independent Cosmetic Manufacturers and Distributors (ICMAD), an industry association representing smaller cosmetic companies.

We are here today to support the Committee's efforts to modernize FDA's statutory authority over cosmetic products.

First, let me briefly describe the Personal Care Products Council and the United States cosmetic industry. Founded in 1894 and based in Washington,

---

* This is an edited, reformatted and augmented version of testimony given on March 27, 2012 before the House Committee on Energy and Commerce, Subcommittee on Health.

D.C., the Council represents over 600 member companies. Council members include such well-known United States and global brands as L'Oreal, Procter & Gamble, Mary Kay, Avon, Johnson & Johnson Consumer Companies, Inc., Revlon, Unilever, and Estee Lauder. The Council also includes more than 500 small businesses, who have 50 or fewer employees and an annual revenue under $10 million.

The American cosmetic industry has an estimated $60 billion in annual retail sales, and employs 8.5 million people, directly and indirectly, in the United States. This industry is a net product exporter. It is innovative and entrepreneurial. The industry launches over 2,000 new products every year. Over 90 percent of cosmetic companies are small businesses that have 50 or fewer employees.

We are here today to discuss future FDA regulation of the cosmetic industry. I will make three points:

1) Current FDA regulation of cosmetics, in partnership with strong industry investment in product safety, assures that cosmetic products in the marketplace today do not present a risk of significant illness or injury. Cosmetics are the safest products that FDA regulates.
2) Globalization of the marketplace for these products, together with new technologies and demand for transparency from consumers, support modernization of FDA statutory authority over cosmetics.
3) Continued consumer protection, innovation and growth in the cosmetic industry, will be achieved through strong FDA regulatory leadership and national enforcement of requirements for ingredient and product safety that apply uniformly through the country.

First, the Federal Food, Drug, and Cosmetic Act (FD&C Act) of 1938 creates a strong framework for FDA regulation of cosmetics. Under this law, it is a crime to market an unsafe or mislabeled cosmetic. Under FDA regulations, cosmetic companies are responsible for substantiating the safety of their products, and each of the individual ingredients, before marketing to the public. FDA has the responsibility to provide regulatory oversight through the creation and enforcement of safety and labeling requirements that hold industry accountable and to conduct postmarket surveillance to determine whether a cosmetic is in violation of these requirements. FDA collects samples for examination and analysis as part of its plant inspections and conducts follow-up inspections to investigate complaints of adverse reactions.

Cosmetic products imported into the United States are subject to the same substantive standards as those produced here. They face an even higher regulatory threshold upon entry into the country, because even the "appearance" of adulteration or misbranding subjects them to detention at the border. All labeling and packaging must be in compliance with United States regulations.

The mandate of product safety is not just a matter of law for our members. It is a commitment for each of them and for our trade association. Our companies invest substantial resources in scientific research and safety processes, and work diligently with thousands of expert chemists, toxicologists, dermatologists, microbiologists and other scientific experts to evaluate the safety of cosmetic products before they are marketed. In fact, cosmetic companies have published thousands of studies on new or enhanced safety assessment methods in scientific journals and often lead adoption of these new approaches by regulatory agencies and scientific groups around the world.

Second, like many industries, the cosmetic industry continues to be affected by rapid globalization of supply chains, expansion in foreign markets, new technology, and increased consumer interest in product information. In much the same way that market changes require companies to adjust business plans, these global challenges justify the modernization of regulatory structures.

The basic statutory provisions that govern FDA regulatory authority over cosmetics today were put in place in 1938. Since 1938, FDA and the cosmetic industry have worked together to keep pace with changing technology by promulgation of creative regulations and the establishment of new regulatory programs. FDA issued regulations requiring safety substantiation of all cosmetic products and ingredients prior to marketing. Based on industry petitions, FDA established programs for the registration of cosmetic manufacturing establishments, the listing of cosmetic products and ingredients, and the submission of adverse reaction reports. At the request of FDA, industry established the Cosmetic Ingredient Review under which the safety of cosmetic ingredients is reviewed by independent expert academic scientists. These are only a few examples of the many FDA and cosmetic industry collaborations to assure product safety. But even though FDA has repeatedly stated that cosmetics are the safest products they regulate, it is time to bring FDA's statutory authority up to date.

Third, we believe that Congress can address these developments by making simple but important changes in the statutory authority over cosmetics.

We offer the following 7 principles to guide this effort. We support enactment of legislation that includes all of them.

1) Enacting into law the existing FDA programs for registration of manufacturing establishments and listing of cosmetic products.
2) Requiring submission of reports on adverse reactions that are serious and unexpected.
3) Mandating FDA regulations establishing good manufacturing practices for cosmetics.
4) Establishing programs to require FDA to review and determine whether controversial cosmetic ingredients and constituents are or are not safe, followed by strong FDA enforcement.
5) Requiring FDA review of all Cosmetic Ingredient Review determinations on cosmetic ingredient safety and either acceptance or rejection of those determinations, followed by strong FDA enforcement.
6) FDA establishment of a national cosmetic regulatory databank for use by everyone.
7) An unambiguous Congressional determination that, as modernized, the revised statute will apply uniformly through the country.

Concerns about cosmetic ingredient safety must be addressed as rapidly as possible by FDA scientists, who can then advise consumers about the safety of products they use every day. We believe Congress should enact a statute that defines a clear path for any person, organization, state or local official, or company, to request that FDA review the safety of a cosmetic ingredient or constituent and make their findings public in an enforceable specified time period. We believe this will allow concerns about cosmetic ingredients and constituents to be resolved expeditiously by the appropriate expert federal agency -- FDA.

It is essential in this legislation that FDA's regulatory authority over cosmetics is firmly established as comprehensive and paramount. It is extremely important for the vitality of the industry that FDA establish national standards on safety that apply in every state. It is impossible to formulate innovative products if different safety standards apply in different states. And FDA's authority is undermined if states create regulatory régimes for cosmetics that are different from FDA regulation of cosmetics. That is why national uniformity of these regulatory changes is critical to our support of this legislation.

Chairman Pitts, and Ranking Member Pallone and Members of the Committee, thank you again for the opportunity to present our proposal. We look forward to working with you on this matter.

In: Cosmetics and FDA Regulation
Editors: A. Garcia and R. DiBartolo
ISBN: 978-1-62257-892-4

Chapter 7

TESTIMONY OF CURRAN DANDURLAND, CO-FOUNDER AND CEO, JACK BLACK SKINCARE. HEARING ON "EXAMINING THE CURRENT STATE OF COSMETICS"*

Good morning Chairman Pitts and Ranking Member Pallone, my name is Curran Dandurand. I am the Chief Executive Officer of Jack Black LLC., a Company I founded twelve years ago with my husband Jeff Dandurand and my colleague Emily Dalton. We founded the company with our combined life savings and a vision of a market segment that we believed was under served. The Company when formed was a private company and remains so today.

Our Company, Jack Black LLC is headquartered in Carrollton, Texas. We develop and market quality personal care products for men under the Jack Black brand name. Our Jack Black line includes skin care, shaving, sun protection, body care, hair care and fine fragrance products. Through development of premium quality, innovative products along with our market positioning, we have been able to grow and expand the Jack Black line from the original 12 products we launched in 2000 to over 50 products which are currently in the line today. Jack Black is sold in all 50 states in the United States. Our Retailers include Neiman Marcus, Nordstrom, Saks Fifth Avenue, Bloomingdales, Sephora, Ulta, AAFES and over 500 independent specialty stores, resorts and spas. We are also distribute our products outside the U.S. in Canada, Mexico, the UK and other international markets. While I have

* This is an edited, reformatted and augmented version of testimony given on March 27, 2012 before the House Committee on Energy and Commerce, Subcommittee on Health.

brought sample products with me which you will see in front of me this morning, you can see our full line of products at www.Getjackblack.com. Please note that there is no connection between our Company and the actor Jack Black.

When we started our Company it was just the three of us and we operated out of our homes. We now employ 39 people and we have office and warehouse facilities. For manufacturing we still rely on independent U.S. based cosmetic manufacturers, who manufacture and fill our products. These Companies also assist us in the development process of new products for our line. We source virtually all of our packaging domestically and have instructed our suppliers to source product packaging from U.S. produced packaging materials when possible. As U.S. entrepreneurs we remain committed both directly and indirectly in our manufacturing and sourcing activities, to ensure that we support U.S. jobs and economic growth in the U.S.

Prior to founding Jack Black, I had the privilege of working for Mary Kay Inc. for 17 years. I served in a variety of positions with ever increasing responsibility. I started in an entry level market research position and was promoted to various senior level marketing positions, including Executive Vice President of Global Marketing and Business Development. I was responsible for worldwide marketing programs, brand strategy, and product development for company operations in 35 countries around the world. During my tenure as head of Global Marketing, Mary Kay's worldwide sales more than doubled. I started my career as assistant buyer at Neiman Marcus in Dallas, Texas.

I graduated summa cum laude from Vanderbilt University in Nashville, Tennessee with a political science major, and received my Masters of Business Administration from Southern Methodist University in Dallas, Texas. I currently reside with my husband and partner Jeff and our two children in Dallas, Texas.

I am here today as a small business owner. I am also a member of the Independent Cosmetics Manufacturers and Distributors Association, commonly referred to as ICMAD. ICMAD is a nonprofit trade association that was founded 38 years ago to provide educational programs and services to assist the small to midsized companies, and to help them succeed in the rapidly changing, highly competitive cosmetics and personal care industries. ICMAD currently has over 650 member companies. ICMAD provides a series of educational and training events to assist its members in understanding and complying with the laws and regulations which govern cosmetic and personal care products. These programs enable ICMAD members to better understand industry best practices in manufacturing and safety standards, as applied to all aspects of developing, manufacturing, distributing and selling cosmetic and personal care products. Since 1983 ICMAD has sponsored educational events at which representatives from FDA office of Cosmetics and Colors and CDER have educated members and nonmembers alike on the FDA's cosmetic and OTC programs including its voluntary cosmetics registration program. My Company is a member of ICMAD and I have been a Director on the ICMAD Board for the last three years.

When we started our business there were only a limited number of companies that marketed a full line of personal care products for men. Today and in part due to our own success, this has changed with many more brands in this category. Some of these brands are being marketed by large multinational companies with significant advertising and marketing resources. For smaller companies like ours that don't have these resources, the key to growth is product innovation and consistent product quality. We have to make sure that we continue to offer new, effective and exciting products that are consistent with our core brand values and positioning.

Product safety is a key part of our brand values. The first step in our innovation process is to make certain that the ingredients we propose for use in any new product formula are safe. We have all of our proposed ingredients reviewed by experts in the field of ingredient safety for topically applied personal care products. Our experts review the scientific literature on the ingredients, along with their experience with the ingredients, to confirm that such ingredients are safe for use in personal care products. Once the ingredient safety is confirmed we then confirm that the combination of ingredients proposed for use in the product formulation is also safe. Consistent with industry standards, all of our proposed formulations are tested using the Human Repeat Insult Patch Test (HRIPT) methodology to ensure that the formulation as a whole is nonirritating and non-allergenic. All of our HRIPT studies are conducted under the direction of and reviewed by a dermatologist. Once our products are fully tested we then proceed to consumer panel testing to confirm product performance and consumer acceptance.

The other key concern in product development is making certain that our products can be produced under our costing criteria and that they are fully compliant with the laws of all jurisdictions in which the product will be marketed. Currently within the United States there has been a movement to create separate state requirements. These regulations would be separate and apart from, and inconsistent with, the federal standards established by the FDA. Compliance with separate state laws that are inconsistent with federal standards would necessitate labeling changes, reformulation, excess packaging and extensive registration requirements, which are simply not feasible for small companies like ours, even successful ones. Smaller companies cannot afford to carry separate inventories to meet the different state requirements; and cannot afford the regulatory staff needed to meet the registration requirements contained in some of the proposed state legislative initiatives. Having to cope with potentially fifty different standards on labeling, ingredient safety and registration would be impossible for a small company.

The science does not change from state to state therefore it does not make any sense from the standpoint of simple logic for there to be varying state regulations regarding cosmetics regulations and safety standards.

Myriad diverse state regulations would substantially increase the cost of producing and distributing personal care products, with a disproportionate impact on smaller companies. This would then lead to small companies either going out of business due to the high cost of compliance, or having to pull out of doing business in those states with costly, onerous regulations and/or dramatic increases in the price of the products without improving the safety or quality for the consumer.

For the benefit of all stakeholders, consumers, personal care marketers as well as regulators, there needs to be one consistent national standard which protects consumer health and safety and provides clear direction and certainty for the regulated companies and the regulators. This would mean transparency in all health and safety decisions and a single forum where all can participate. We support the modernization of the FDA laws that creates a National Standard for cosmetics. I believe this will best protect the health and safety of our consumers and provide a strong foundation for growth and success of our small entrepreneurial companies that create jobs here in the U.S.

Thank you for providing me the opportunity to appear before you. I would be happy to answer any questions you may have.

In: Cosmetics and FDA Regulation
Editors: A. Garcia and R. DiBartolo
ISBN: 978-1-62257-892-4

Chapter 8

TESTIMONY OF DEBORAH MAY, PRESIDENT AND CEO, WHOLESALE SUPPLIES PLUS, INC. HEARING ON "EXAMINING THE CURRENT STATE OF COSMETICS"*

Good morning Mr. Chairmen Pitts, Ranking Members Waxman and Pallone, and Members of the Subcommittee on Health. Thank you for this opportunity today. My name is Deborah May and I am President of Wholesale Supplies Plus in Broadview Heights, OH.

I am honored to offer testimony on behalf of the handcrafted soap and cosmetic industry. These small and micro businesses produce quality, customized products. With more than 200,000 such companies nationwide, they make significant economic contributions in communities throughout the country. My hope is that as the Subcommittee moves ahead with legislation to improve cosmetic safety, it will include provisions that recognize the products and contributions of the handmade soap and cosmetic industry.

I became a handcrafted soapmaker 16 years ago, not by choice rather by necessity. At the time, I was a Registered Nurse in the ICU at The Cleveland Clinic. But on August 1, 1996 I gave birth to my second daughter who was diagnosed with cortically blindness and severe autism.

---

* This is an edited, reformatted and augmented version of testimony given on March 27, 2012 before the House Committee on Energy and Commerce, Subcommittee on Health.

In the months that followed, I lost my job because of my daughter's around the clock medical care. Our bills became overwhelming. My husband, a Catholic high school teacher, and I were drowning in debt. Our secure, predictable middle class life was gone. I sought support through online forums with other mothers facing similar challenges. Through one exchange, I was introduced to the art of making handmade soaps and lotions. I found a handcrafted product forum online and these women taught me how to make products safely, comply with ingredient labeling laws and answered all of my questions. I was amazed how easy it was to make small batches of handmade cosmetics.

I began to give my products out to friends and family. Before I knew it they were encouraging me to sell my products for profit. Although fearful, I took a deep breath and registered for a local high school craft show. People loved the products and I went home with empty crates and a cash box full of money. After that, I registered for every craft show I could find.

Caring for my daughter was my first priority, but I had found a business that allowed me to do both. I landed my first wholesale account through a customer whose brother owned a shop in California. He was delighted I would make 10 bars of custom soap in any combination of scent and color and fill the order within 48 hours.

At home, I built my business, and it worked. I loved what I was doing, and most important, it saved my family from foreclosure and allowed us to begin to pay the overwhelming medical bills.

The following year, I began teaching adult classes on handcrafted soap and cosmetic making. Families were financially hurting and many were looking for a way to make ends meet. In 1999, I founded the company Wholesale Supplies Plus. My goal was to teach others how to make their own handmade cosmetics and provide supplies in quantities and sizes micro-businesses could afford.

Today, Wholesale Supplies Plus is one of the leading ingredient suppliers for very small businesses producing handmade soaps, lotions, bath salts and other topical cosmetics. Since 2010, my company has serviced over 80,000 unique businesses in the United States. We are on target to exceed $10 million in sales this year and have 35 employees.

I wanted to share my personal story of how I began my handcrafted products business, because it is not all that different from most people who are hand producing soaps, lotions and cosmetics.

Recently, handmade industry leaders pooled data that confirms the industry is over 200,000 small businesses hand producing small batches of soaps and cosmetics. Ninety-five (95) percent are woman-owned and average between 1 to 3 employees that translates to between 200,000 and 600,000 jobs in the U.S. These small businesses help families and retirees pay mortgages, rent, food and household bills.

The handmade cosmetic industry supports Congress' efforts to ensure safe cosmetics, and we believe our products are of the safest on the market. Our ingredients support this claim, as 95% of what is used by hand-made cosmetic companies is food-grade products found in grocery stores. The remaining 5% are natural essential oils and synthetic chemicals currently deemed safe when used as directed by the ingredient manufacturers. Handcrafted soap and cosmetic makers are not splitting molecules to make new ingredients or traveling to the rainforest to find new plants that prevent wrinkles. Sugar Scrub, a best seller, contains food-grade olive oil and sugar with an aroma such as lavender oil.

The handcrafted soap and cosmetic industry support the principle of identifying ingredients of concern. If the FDA determines an ingredient is unsafe, we don't want it in the products our family uses and won't sell it to our customers.

We support the principle of the giving the FDA recall authority for cosmetics. Frankly, I imagine most consumers believe the FDA already has that authority.

We support the principle of requiring adverse event reporting of serious reactions that cause loss of life and/or hospitalization.

We support the closing of labeling loopholes such as the current incidental ingredient exclusion that is used to hide such things as preservatives from the consumer. If a product label reads "preservative free," consumers should have confidence that there are no preservatives.

We support small business exemptions for facility registration allowing small and micro businesses to make products for themselves, friends and family without the fear of breaking federal laws. Small business exemptions are vital to the handmade product industry – to encourage entrepreneurial growth and create local jobs.

We support small business exemptions for fees. Registration fees will be a barrier for entering the market and will shutdown all but a few of the 200,000 companies now in the handmade industry. For growing, established businesses, I urge the Subcommittee to consider a sliding scale. A company selling $2 million in products should not have to pay the same fee as a company selling $100 million.

We do not support a requirement to register with the FDA individual product batches or requiring the producer to register each ingredient supplier used in that batch. The handmade cosmetic industry makes very small, custom order batches. We may make 50 jars of different sugar scrubs several times a week and buy sugar and olive oil from several different grocery stores or food warehouses. Under the considered provision of notifying the FDA of a change in suppliers, it presumes truckload bulk purchases of ingredients. That is not the case with small businesses. We buy as needed and it fits in a shopping cart.

Emerging small businesses grow by making and marketing products. If legislation is written in such a way that it strengthens the standing of safe ingredients then the volumes of paperwork for batch reporting serves only to give large corporations, that buy in truckloads and produce millions of units in a single batch, an even greater market advantage. Quite simply, in one month if a small business were to make 100 batches of 10 differently scented sugar scrubs using 10 different sugar suppliers, the reporting requirement would result in a minimum of 1000 reports for just one product. If soaps and lotions are included, the business is easily looking at 5000 reports in a 30-day period.

I am not here to seek exemptions for Wholesale Supplies Plus or companies like mine that have had the good fortune to grow. I am here so that the 200,000 small businesses making handcrafted cosmetics have the same opportunity for growth and the chance to become the next success story...like Bert's Bees, Mary Kay Cosmetics and even James Gamble of Proctor &, Gamble -- all of whom started as handcrafted micro businesses.

As President Ronald Reagan said during his first inaugural address, "government can and must provide opportunity, not smother it: foster productivity, not stifle it." On behalf of the handcrafted soap and cosmetic industry, I hope to work with the Subcommittee to enhance cosmetic safety while fostering opportunity and growth for small companies.

Testifying today has been an honor and a privilege. Thank you.

Small Handmade Cosmetic Manufacturers (by state)

State	Small Businesses By State	State	Small Businesses By State
APO (Military Address)	44	MS	1,874
AK	350	MT	982
AL	2,766	NC	6,898
AR	1,844	ND	466
AZ	3,970	NE	1,052
CA	19,086	NH	1,490
CO	4,480	NJ	5,080

State	Small Businesses By State	State	Small Businesses By State
CT	2,296	NM	1,368
DC	420	NV	1,552
DE	588	NY	11,692
FL	11,738	OH	11,874
GA	6,542	OK	2,632
HI	624	OR	3,228
IA	2,028	PA	9,558
ID	1,352	RI	582
IL	7,528	SC	2,340
IN	4,592	SD	488
KS	1,932	TN	4,526
KY	2,922	TX	13,172
LA	3,480	UT	1,950
MA	4,472	VA	5,334
MD	4,032	VT	744
ME	1,756	WA	5,706
MI	8,736	WI	4,644
MN	4,156	WV	1,464
MO	4,660	WY	584
Column Total	106,394	Column Total	101,280
Grand Total of Both Columns			207,674

Data is an estimate based on cumulative data shared by:
Wholesale Supplies Plus, Inc. Bramble Berry, Inc.
The Handcrafted Soapmakers Guild

In: Cosmetics and FDA Regulation
Editors: A. Garcia and R. DiBartolo
ISBN: 978-1-62257-892-4

Chapter 9

TESTIMONY OF DR. MICHAEL DIBARTOLOMEIS, CHIEF OF THE SAFE COSMETICS PROGRAM, CALIFORNIA DEPARTMENT OF PUBLIC HEALTH. HEARING ON "EXAMINING THE CURRENT STATE OF COSMETICS"*

Good morning Mr. Chairman and distinguished members of the Energy and Commerce Health Subcommittee. My name is Michael DiBartolomeis and I am chief of the Safe Cosmetics Program in the California Department of Public Health. I earned a PhD in toxicology from the University of Wisconsin in 1984, with additional formal education and training in biochemistry, molecular biology, epidemiology, and public health. I am certified by the American Board of Toxicology and have presented original research in over 270 publications, conference proceedings, and government reports. For more than 28 years, 23 in state government, I have worked in environmental and occupational health, health risk assessment, laboratory research, and chemical policy development.

As chief of the California Safe Cosmetics Program, which was established in 2006 and is the first state cosmetics-regulatory program in the nation, I believe I offer a unique perspective on the safety of cosmetic products and the

* This is an edited, reformatted and augmented version of testimony given on March 27, 2012 before the House Committee on Energy and Commerce, Subcommittee on Health.

challenges in adequately protecting consumers. In my testimony I will briefly address:

1) growing public concern about the safety of cosmetic products;
2) challenges in evaluating cosmetic product safety;
3) benefits of the California Safe Cosmetics Act of 2005; and
4) five elements that I believe would assist in the evaluation of the safety of cosmetics and protecting public health.

First, why is there growing concern with regard to the safety of cosmetics products? During my six-year tenure directing the California Safe Cosmetics Program, I have heard concerns from many consumers and professionals in the personal care industry about:

- the negative effects cosmetic products might have on infants, children, the developing fetus and other susceptible persons, such as salon workers who are consistently exposed to greater amounts of certain cosmetic products;
- the lack of information available on critical cosmetic product ingredients, such as fragrances, and the weak labeling laws for professional-use products;
- the number of chemicals and formulations on the market that have not undergone toxicity testing; a problem commonly referred to as "data gaps;"
- the unknown impacts on cosmetics users' health from long-term, low-dose exposure to individual chemicals or chemical mixtures; and
- insufficient consumer and workplace safety standards and enforcement.

Cosmetics are any product sold or marketed with the intent that they be applied to any part of the human body for cleansing, beautifying, promoting attractiveness, or otherwise altering the appearance of a person. We use cosmetics from the time of infancy, or even *in utero*, through our senior years on a continuous, daily basis. Exposure to chemicals in cosmetics can occur from breathing vapors or particles, inadvertent swallowing, and of course from applying them to the skin and eyes. Women use an average of 15 cosmetic products per day, and daily usage may be as high as 50 products, according to women surveyed in a 2011 Portland State University study. Many might find this statistic startling because they do not understand that the universe of cosmetic products goes well beyond lipstick and eye shadow; it includes everything from toothpaste to shampoos to deodorants to shaving cream and even sunscreens.

Although we have known for decades about air and water pollution, in the past 12 years we have also found that people's bodies are biological reservoirs for environmental chemicals. In studies published by the Centers for Disease Control and Prevention and other agencies and academic researchers, it has been reported that more than 200 chemical residues or metabolites from environmental sources are present in people's blood, urine, and breast milk and in the cord blood of newborn babies. Some of these chemicals are ingredients or contaminants in cosmetic products such as the plasticizers called phthalates, phenols such as bisphenol-A and benzophenone, hormone-mimicking chemicals such as synthetic estrogens and parabens, volatile organic compounds like toluene, and heavy metals such as lead and mercury. None of these chemical residues in our bodies serves any beneficial physiological purpose.

Second, what are some of the challenges we encounter when assessing the safety of cosmetic products and protecting public health?

The cosmetics provision within the Federal Food, Drug and Cosmetic Act was written in 1938 and has not been significantly amended in over 70 years. Since that time, the cosmetics industry has grown to be a multi-billion dollar industry with products being marketed world-wide and sold not only in retail stores but by individuals working out of their homes and over the Internet. While the industry has changed, the provisions in the federal law for regulating cosmetics have not. As a result:

- the law requires government to show harm before a cosmetic product can be taken off the market; in other words, the burden of proof falls on the government.
- the law does not require safety testing of cosmetics before they are marketed and therefore products that might not have been evaluated for safety, especially for repeated exposures over a person's lifetime or during pregnancy, may be lawfully sold.
- cosmetic labels are not required to disclose some ingredients, most notably fragrances, colors, and flavors; and except in very limited instances, professional salon product labels do not need to list any ingredients and there are no requirements for disclosure to the federal government of ingredient lists for cosmetic products.

- while manufacturers may have inherent incentives to test for immediate and obvious harmful effects of their cosmetic products, for example, allergic reactions, rashes, or chemical burns, they have almost no incentive to test products for their potential to cause serious latent harms, such as cancer, where it will be difficult if not impossible for consumers to prove the source of their illness.
- chemicals that cause cancer, reproductive and or developmental harm, and other chemicals such as those that disrupt the endocrine system, are consistently ending up in cosmetic products.

Third, what is the California Safe Cosmetics Act, and why is it necessary?

The California Safe Cosmetics Act was signed into law in 2005, and is based on the principle of "Right-to-Know." The Act requires manufacturers with aggregate sales of greater than $1 million and whose products are sold in California to disclose to the State all intentionally added chemical ingredients in their products that are known or suspected to cause cancer or reproductive and or developmental toxicity, regardless of the concentration of the chemical. To facilitate this, the Program launched a unique electronic reporting system in 2009, which the industry helped to design.

Although the Safe Cosmetics Act does not set product safety standards or ban any products, it responds to public concerns about the safety of cosmetics by empowering them to avoid the most toxic chemicals, and it thereby also promotes product reformulation.

The Act grants authority to the State's Safe Cosmetics Program to conduct audits, investigations, and health-based studies, and requires manufacturers to submit any additional information on their products as deemed necessary by the Program for conducting these assessments. Note that FDA does not have comparable authority. The Program is required to inform regulatory authorities in the State when its investigations reveal a public or occupational health concern.

At the end of last year, 17,060 unique cosmetic products were reported to the Program as containing one or more chemical ingredient known or suspected to be carcinogens or reproductive or developmental toxicants, as reported by 700 unique companies. In total, 24,664 hazardous ingredients were reported in these products, represented by 96 unique chemicals. How has the California Safe Cosmetics Act benefited public health?

First, the data collected by the Safe Cosmetics Program has been accessed by governmental agencies and other organizations and used to support laboratory analyses of cosmetics such as nail polishes and removers, shampoos for infants and children, and women's make-up. From these efforts, health advisories and guidance are developed to aid the consumer in understanding the risks and benefits from using certain cosmetic products in order to make healthy choices when shopping.

Second, in the past two years, the Program has initiated its own public health investigations of specific cosmetic products that contain reportable chemicals under the Act. Some of these investigations, such as skin lightening creams that contain mercury are ongoing, and I cannot describe them here in detail. However, to illustrate how the Act can be used to benefit public health, I will give one example.

In March of 2010, the Program started receiving phone calls from professional hair stylists and clients complaining about health effects from using a hair-straightening product called Brazilian Blowout. Complaints included burning eyes, nose, throat, scalp; hair loss; asthma episodes; skin blisters; and other effects consistent with a class of volatile chemicals called aldehydes. Historically, these hair-straightening products have contained formaldehyde, a known human carcinogen, as a key active ingredient. However, this product was being advertised as "formaldehyde-free." We noted at the time that the manufacturer of this product did not report to the State that its product contained formaldehyde, even though at least one other similar product had been reported by another manufacturer as containing formaldehyde. What happened over the course of the next 22 months is too long a story for me to tell. However, the end result is informative:

- On January 30, 2012, California announced a settlement with the makers of Brazilian Blowout, requiring that they warn consumers about the dangers of using this product and stop falsely advertising and marketing their product as formaldehyde-free. In addition, they were required to report their product to the State as containing formaldehyde, update the material safety data sheets required for industrial products, and pay a fine.
- In its press release, the California Department of Justice stated, "Today's settlement is the first government enforceable action in the United States to address the exposures to formaldehyde gas associated with Brazilian Blowout products. It is also the first law enforcement action under California's Safe Cosmetics Act, a right-to-know law enacted in 2005."
- Despite efforts to call attention to the dangers of using hair straightening products containing formaldehyde, these products are

still being used on a daily basis in salons across the United States. In contrast, six countries have recalled the use of formaldehyde-based straighteners, including Canada, France, and Ireland.

- On March 6, 2012, the New York Times reported that the makers of Brazilian Blowout agreed to settle a class-action lawsuit for $4.5 million. The Chief Executive Officer said the settlement will be paid by his insurance company and was quoted saying: "We get to sell the product forever without reformulation ... that's the acquittal we've been waiting for."

Although the sale of this product in California violated five separate state health, environmental, and consumer laws and resulted in numerous acute injuries, we have not to date been able to get it off the market. The best we could do was to require warnings and other restrictions that would reduce the product's market appeal and increase the level of precaution exercised by product users.

Finally, in my capacity as the Chief of the Safe Cosmetics Program I have had the opportunity over the past six years to contemplate the challenges with regard to evaluating cosmetic and other consumer product safety and I have arrived at five elements, which I believe would help in evaluating the safety of cosmetics and protecting public health:

1) Reverse the burden of proof from the government having to demonstrate cosmetic harm to the manufacturers having to document product safety, through pre-market safety testing of new cosmetic products using a tiered battery of toxicity tests. That is, start with inexpensive screening level tests and then, depending on the results, move onto more complex tests if needed.
2) Ensure that toxicity testing data, safety data, and other key information is available to government agencies and to consumers.
3) Improve cosmetics labeling so that all chemical ingredients, including fragrances, colors, and flavors for any cosmetic, including professional-grade products, are disclosed to consumers.
4) Establish safety standards for cosmetic products and issue prompt mandatory recalls of cosmetics that have been found to be unsafe, adulterated, or misbranded.
5) If a standing science advisory committee for cosmetic safety is thought to be valuable, require that committee members have no

conflicts of interest, and that the committee be wholly independent rather than industry-sponsored.

In closing, I want to say that in my role as the Chief of the Safe Cosmetics Program, I have personally attended meetings where dozens of people have told their stories of illness and expressed their concern about the safety of using cosmetic products at work or at home. Afterward, I go back to my office and I ask myself how I can make the California Safe Cosmetics Program work better to inform policy-makers and the general public about the data gaps regarding cosmetic product safety. I don't know how many cases like Brazilian Blowout exist. However, the fact is, cosmetic products that contain known human carcinogens or chemicals that impair human reproduction or development are marketed and sold, without adequate safety testing, because the existing law allows it. This is a serious public health problem, which we can prevent because there are some very workable solutions to consider.

I want to thank the committee for inviting me to testify and I would be happy to answer any questions you might have for me.

INDEX

A

B

C

D

E

F

G

H

I

J

K

P

Q

R

W

Z